PATHWAYS TO THE PLANETS

Memoirs of an Astrophysicist

By

John R. Strand

1663 Liberty Drive, Suite 200
Bloomington, Indiana 47403
(800) 839-8640
www.AuthorHouse.com

© 2004 John R. Strand.
All Rights Reserved.

No part of this book may be reproduced, stored in a retrieval system, or transmitted by any means without the written permission of the author.

First published by AuthorHouse 10/18/04

ISBN: 1-4184-9683-9 (sc)

Library of Congress Control Number: 2004098053

Printed in the United States of America
Bloomington, Indiana

This book is printed on acid-free paper.

PREFACE

This book will become an important resource for those readers interested in the history of aerospace, astrophysics, computer science and mathematics. It is a true, first hand account of a computer system, which provided the information necessary to guide our spacecrafts to the planets. This represented a unique scientific odyssey and a great leap in understanding the physics of our solar system. It chronicles scientific breakthroughs relating analysis from the world's greatest physicists (Galileo, Kepler, Newton and Einstein), to our modern astro-navigation systems. This historical thread is traced through the uncertainties of the time of Sputnik to the glorious successes of Mariner, Viking and Voyager. From pictures of the planets on every school house wall to global positioning and continental drift, this is history of the personal accomplishments of a small handful of scientists at the Jet Propulsion Laboratory in Pasadena California. With hardly any distinction, they left the world an unparalleled footprint at the highest levels of scientific achievement.

As a leading mathematician and computer programmer involved with these projects, Mr. Strand paints a picture of a complex technology with its inter-woven success and frustrations insightfully colored by witty stories. He also describes the twists and turns of his own problem-solving experiences from childhood to the present day. This very human setting unfolds into a surprisingly diverse world of scientific experiences. Chemistry, geo-physics, electronics and fiber optics are some of the additional topics. Non-technical readers will find a rich and surprising look into human interactions through Mr. Strand's amusing anecdotes.

Linnea Rodes Strand

NOTE FROM THE AUTHOR:

I have interacted with many different personalities over the years and found virtually everyone that I met fascinating. It is not my intention to offend where I portray a contentious personal encounter. These are necessary literary devices that tell the complete story. I feel in most of these cases a heroic individual eventually emerges. Some of the names have been changed so that I may be certain no one will be offended should they read about themselves. The navigation system we worked on was an enormous project. A large number of scientists were involved, I did not work with everyone. I hope future historians will be able to extend my writing to include their stories.

Another important point is that people can be funny, but drugs are not. So please understand that my references to marijuana are not at all intended to promote drug use.

TABLE OF CONTENTS

PREFACE v
NOTE FROM THE AUTHOR: vii
INTRODUCTION xv

CHAPTER 1
   THE EARLY YEARS 1

CHAPTER 2
   WALKING SKYSCRAPER 13

CHAPTER 3
   AS THE CROW FLIES 17

CHAPTER 4
   REFLECTIONS ON AVERY LABEL 25

CHAPTER 5
   NORTH AMERICAN APOLLO 31

CHAPTER 6
   STRING OF PEARLS 37

CHAPTER 7
   JPL HISTORY OF DPODP-DPTRAJ, 1966-1971 43

CHAPTER 8
   THE LAST MARINER PARTY 57

CHAPTER 9
   UNIVAC VS. IBM 61

CHAPTER 10
   JPL FRIENDS (DEATH VALLEY OR BUST) 73

CHAPTER 11
   IBM VS. INFORMATICS 77

CHAPTER 12
   RICH BRODIE, INVENTOR 81

CHAPTER 13
   DRS. LAWSON & KROGH..85

CHAPTER 14
    JOHN EKELUND 89

CHAPTER 15
    INSTANTANEOUS MANEUVER 93

CHAPTER 16
    DAVE HILT 97

CHAPTER 17
    ALVA JOSEPH 99

CHAPTER 18
    XX NEWHALL, AKA SKIP NEWHALL 103

CHAPTER 19
    TED PAVLOVITCH 107

CHAPTER 20
    AHMAD KHATIB 111

CHAPTER 21
    TED MOYER 115

CHAPTER 22
    FRAN STURMS 119

CHAPTER 23
    DAN ALDERSON 123

CHAPTER 24
    PH.D D D D'S 125

CHAPTER 25
    LET'S CALL HIM DR. ABBOT 129

CHAPTER 26
    BEER AND MORE BEER 133

CHAPTER 27
    BAMBI VS. GODZILLA 135

CHAPTER 28
    DUMB DOWN 137

CHAPTER 29 PREDICTS	139
CHAPTER 30 JACOBI SYSTEMS	143
CHAPTER 31 CLAM	147
CHAPTER 32 IT WAS UP TO AZUL	153
CHAPTER 33 CORNING AND REICHERT-MCBAIN	157
EPILOGUE	165
APPENDIX I	167
ATTACHMENT A	177
ATTACHMENT B	179
ATTACHMENT C	181
ATTACHMENT D	183
ATTACHMENT E	185
APPENDIX II	187
APPENDIX III	193
APPENDIX IV	197
ACKNOWLEDGMENTS	199
INDEX	201

Viking Launch
KSC 8-20/1975

INTRODUCTION

The mathematical processes used in monitoring the motion of celestial objects

Navigation of manned or unmanned spacecraft within the solar system is a relatively recent development in the sphere of human endeavor. I was born before serious national goals were established for the exploration of outer space. This process started fewer than fifty years ago when tiny star-like objects were seen traversing the night sky.

Von Karman, Goddard and Von Braun are pioneers we commonly associate with early rocketry. The advent of space exploration is not, however, attributed to any individual scientist but rather to a small satellite named "Sputnik," launched in 1957 by the Soviet Union.

Early low orbit satellites launched by the United States and Russia required relatively simple on-board guidance to approach escape velocity and orbit the earth. Complex objectives promulgated by the Defense Department, the communications industry, and National Aeronautics and Space Agency (NASA) for Apollo and planetary exploration dramatically increased the need for more accurate navigation. Soon computer-linked tracking antennas spanned the globe and ever more sophisticated spacecraft projects created an industrial complex employing hundreds of thousands of people. The United States would dominate this new technology with men landing on the moon before the end of the 1960s.

It is difficult to denote a single branch of science that was not influenced by the space race. Certainly computer science saw gargantuan technological advances. Every few years new generations of computers allowed engineering problems to be solved which were previously considered impossible. Celestial navigation software was utilizing the increased speed and capacity of these early computers. Mission planning, vehicle design and construction were often well underway before the computers that supported them came off the assembly line.

NASA was very well funded and the Jet Propulsion Laboratory (JPL) coordinated much of the planetary exploration. Within a few years, the United States space program was able to send spacecraft to the closer planets. There was worldwide excitement as the rocky surface of Mars was photographed and tested for life. The planets and their moons were photographed from flybys and these pictures became a part of every school curriculum in America. Only Pluto remained unobserved.

How did the navigation systems accomplish all of this? To trace the origins of Orbit Determination from its roots to present day we need to examine science beginning roughly at a time when the Copernican system became widely accepted.

Galileo helped perfect the telescope and also studied the mathematics of falling bodies. Kepler analyzed data measurements from Tycho Brahe and found the planets move around the sun in ellipses. He would determine additional laws about planetary periods and the area swept out by the Sun-planet line. These laws influenced Newton, who connected falling bodies on Earth, like the apple, to heavenly bodies. To produce his law of gravity did require a new mathematics, which we call calculus. Newton's inverse square law explained Kepler's findings.

Newton had utilized mathematics--specifically, differential equations and the physics of mass attraction--to delineate a solution for the relationship between two bodies revolving around a common point. Computation rather than observation could now determine their position and velocity at any time.

His analytical solution (Newton's Method) is for two point masses, and never was extended analytically to include other massive bodies. It is unlikely additional analytical methods for computing planetary motion will ever be discovered, and at any rate are practically unnecessary. More complex solutions for the equations regarding the motion of celestial bodies will most likely continue to require the nuts and bolts of numerical methods. These hack and slash techniques, unlike the elegant pad and pen methods of Newton, require computations that only digital computers with their incredible speed are capable of performing.

Gauss would add remarkably to Newton's analysis in the 1800s with recursive equations that were used to locate a newly discovered asteroid called Ceres. Ceres, lost shortly after discovery because of bad weather, was re-acquired the next year exactly where Gauss predicted. This elevated him in the public eye and also added to the prestige of this pre-eminent division of mathematics.

Newton's work was so respected that by 1900 physicists were suggesting that everything relating to the mechanics of motion had been discovered. There was, however, a small discrepancy in the orbit of Mercury, and speculation existed that massive stars might not allow light to escape their powerful gravity. Rockets and computers were decades away. Airplanes and radio communication were just being invented, and Albert Einstein was having difficulty finding work.

Einstein had dropped out of school, complaining about rote learning and martinet teachers. His mind was reaching for the stars, and shortly his ideas would turn physics upside down. He eventually found a job in a Swiss patent office and in a few months developed the Theory of Relativity. It is a hundred years old and remains the pre-dominant principle describing the physics of the universe.

Newton had provided an excellent model for explaining the influence mass has on motion. The corrections Einstein supplied were in effect small; however, the concepts Einstein used to explain these improvements were revolutionary. Space, time, light, matter and energy would now have to be re-defined. These concepts unleashed a new non-intuitive perception of the universe. Unquestioned and unassailable until then, celestial mechanics was about to be overhauled.

Einstein, as a young theoretician, could not imagine if and when man-made objects would travel in space. Regardless, his General Theory of Relativity related to the interactions of all matter in space, including that launched from earth. Unwittingly his theories about electro-magnetic waves were subsequently utilized for spacecraft navigation. He would not see the consequences of his theory, but years later as the pictures came streaming back from the planets his extraordinary legacy was confirmed.

Einstein's theories were wonderfully penned in an era when electricity was only available to mechanically move machine parts and produce relatively simple arithmetic calculations. His mind was clearly reaching for the planets, but the technology to support his ideas would not be available for decades.

This leads us to a very basic question. How do the complex facilities at JPL function, and what is the mathematical analysis that computers execute to determine the PATHWAYS TO THE PLANETS?

Consider driving down the highway. The steering wheel is what you use to follow the road. Imagine proceeding along a deserted road and moving the steering wheel in single deliberate planned movements. Between each turn of the wheel, thoughtfully plan your next move and hold it stationary until you decide. Try to move the wheel as few times as possible. Gradually as you get the feel of the road you will be able to anticipate your car's natural tendency to drift right or left. Each time you hold the wheel steady and contemplate the next wheel turn, your car will gradually drift to one side. Each time you move the wheel to correct the drift you will be able to compensate a little better for the vagaries of the road. Experience will reduce the number of adjustments, but eliminating all of them will be impossible.

There are parallels between navigating a car and a spacecraft. Like steering a car, space travel requires periodic changes in direction (midcourse motor burns) as well as continuous monitoring of the effects forces have on the spacecraft's motion. It is the physics of everything in the solar system that ultimately determines this motion. In turn, as detectives searching for the facts, we observe this motion and reach conclusions about the physics of our solar system.

Radio signals (telemetry) give us a good idea where a spacecraft is located and its velocity. They are measured in a variety of ways, one of which is Doppler. These electro-magnetic waves are not communicated instantaneously but are delayed by the amount of time it takes for them to travel at the speed of light to a tracking station.

Each telemetry signal received and stored by the Deep Space Net (DSN) at JPL has been influenced by conditions along its path. From distortions near the Earth (mainly the electron density of the ionosphere) to the curvature of space as described by Einstein, these signals have been adulterated. Mathematics will be crucial in reversing these effects so that the precise position and velocity of the spacecraft at the moment of transmission may be determined.

A second aspect of the orbit determination process involves considerably more mathematics than the telemetry. Referred to as "the equations of motion," they represent the physics of the forces affecting the trajectory of the spacecraft. This is the cumulative result of the work of astrophysicists throughout history as they have attempted to mathematically describe the motion of heavenly bodies.

To underscore the dual functions of orbit determination, there is mathematics to replicate our entire mission and there is telemetry that tracks the real time motion of our spacecraft. We hope both of these representations of the trajectory will be ultimately identical. Minimizing the errors that are inevitably present between these similar measures is the controlling mechanism of orbit determination.

Augmenting the equations of motion so as to align them with the telemetry is a mathematical technique referred to as regression analysis. This results in improved values of the parameters formulated and utilized within the equations of motion. This represents an improvement in our knowledge of the physics of the spacecraft's motion, assuring us it will arrive at its intended destination. As a spin-off, it also improves our understanding of the solar system it flies through and increases our knowledge of fundamental physics.

Monitoring trajectory conformity with the anticipated mission goals is the job of a team of scientists located in the Spacecraft Flight Operations Facility (SFOF) building at JPL. They are the heart and soul of the project and monitor all available data connected with the spacecraft. Small forces not understood or described in the equations of motion inexorably lead to the vehicle departing from the nominal (JPL parlance) or acceptable trajectory. This leads to a committee issuing a decision to execute a mid-course correction based on a wide range of issues pertaining to the integrity of the mission. A rocket motor will fire and realign the trajectory to a nominal status.

Like our steering wheel example, we attempt to move the wheel infrequently but safely. The more accurately we point the car, the more infrequently we move the wheel. Each time we move it we are again on an acceptable course. Gradually, however, the unaccounted or secular (JPL parlance) forces cause us to drift away from our nominal trajectory and another mid-course correction is required.

Each maneuver by the rocket motor expends fuel and opens the door for malfunctions. The more accurate the equations of motion, the less midcourse corrections are necessary. Commensurately, less fuel will be consumed. The largest computer programs executing on the largest computers, requiring the most complex mathematics in the world, are necessary to accomplish these missions. Nothing less will do.

The 1950s comic strip character Buck Rogers did not seem to ever have a fuel problem in outer space. If you recall, he could travel wherever and whenever the adventure required. Unfortunately, modern space travel demands adherence to strict fuel conservation strategies. The key to minimizing fuel is accurate navigation.

Within the equations of motion are thousands of parameters such as the mass of planets. Each of these parameters affects the motion of the spacecraft. (See page 45 for a complete list).

Hundreds of equations use these parameters with some of the most sophisticated computations ever penned by mathematicians. Generally there is not a better exposition of the synergetic world of physics and mathematics than orbit determination.

Unlocking the mysteries of celestial motion entwines the great astrophysicists (Galileo, Kepler, Newton, and Einstein) with their modern counterparts: the JPL engineers, analysts and programmers.

Looking again at correlating data for the purposes of navigating in space, the methods of orbit determination permit the physics and telemetry to be closely aligned. Ideally, you might wish that all the forces acting on the spacecraft are so well understood that at any future moment in time the mathematics (the equations of motion) would tell you exactly its position and velocity. The telemetry signals should then be only corroborative.

Approaching perfection in the 21st century, unimaginable accuracy now makes it possible to study continental drift. Even the tidal movement of the earth's surface due to the lunar cycle is detectable. Global positioning in the millimeter range is now achievable, and is a result of JPL's orbit determination programs.

What would Einstein and the other good old boys think of today's aerospace projects? They would have some difficulty recognizing their mathematics as it is structured in the FORTRAN coding. Each physicist's contributions were the building blocks for the next era. I wonder what a conference of historical astrophysicists would be like? Coffee and donuts would probably be untouched. The future and the past were about to be reconciled.

Newton would be surprised to learn that his solution to the two-body problem had not been expanded to include additional bodies. Computers using numerical techniques now dominate celestial mechanics. Additional analytic solutions augmenting Newton's work would be nothing more than a curiosity to him. Interest in pure mathematics has diminished with the spectacular success of computers and their natural affinity for engineering problems. Gauss did understand that iterative analytic techniques would be useful in tracking celestial objects but could not possibly comprehend how electronics would revolutionize mathematics. Only Einstein had a glimpse of early rockets and computers. He passed away two years before Sputnik.

History periodically records sensational leaps in knowledge. These are the bulkheads in my opinion for periods of reconciliation. Modern information technology is a result of consolidations involving computers, lasers, fiber optics and electronics discovered decades earlier. Is another breakthrough likely involving celestial mechanics, or has it possibly been conquered? I asked Galileo, and he said his studies with falling objects introduced the use of mathematics to help explain motion. He was overwhelmed with the widespread use of his telescope. He did say he was gratified to learn that people who believe the world is round are no longer imprisoned and freedom from hypocrisy will be essential for the next scientific breakthrough.

Kepler was surprised to learn how much Newton had accomplished using his discovery of conic motion. He realized the inadequacy of mathematics in his time prevented further discovery. He suggested more access to scientific equipment by the public was the key to the future.

Newton wished he had pursued his musings on relative motion as it applies to rotation. "Newton's bucket" and distant stars were still on his mind and he felt the next scientific thunderclap was out there in front of us; we only needed to pick it up. There were still falling apples all around us.

Einstein was excited to find out how relativity had contributed to orbit determination. Without it, Voyager would have run out of fuel. Although JPL's fundamental coordinate system at the heart of their orbit determination computer program used the Newtonian (XYZ) principle, relativity corrections were formulated for the equations of motion, time keeping and telemetry. Our minds at JPL had not been able to leave the comfort zone of the 19th century and embrace the new exotic relativistic systems. Still, he congratulated us for the interesting innovations our "correctionist" methods had followed.

At the beginning of his work Einstein had noted how his theory explained Mercury's orbit precession. An experiment had revealed the bending of starlight during an eclipse of the sun in 1919, and this first confirmed his theory. He was especially interested to find out that the modest speed of Voyager was a platform for another verification of relativity. It did not rely on velocities approaching the speed of light; rather the lengthy mission (twelve years) and the great distance traveled eventually provided an accumulated measurable perturbation which departed from Newtonian physics.

Einstein emphatically stated future discoveries would come to those who question the old axioms as he had. He appealed for more open discussions in classrooms with instructors unhampered by university politics.

All the great physicists were astonished to learn that hundreds of thousands of people had participated in undertaking the great missions. What a future, they thought, for young people to continue the adventure.

Besides the views of the earth, they seemed to feel Io and the volcanoes represented the most compelling pictures from space. Here was dynamics, here was excitement, and this alone would have made it all worthwhile!

"Marble Gang"

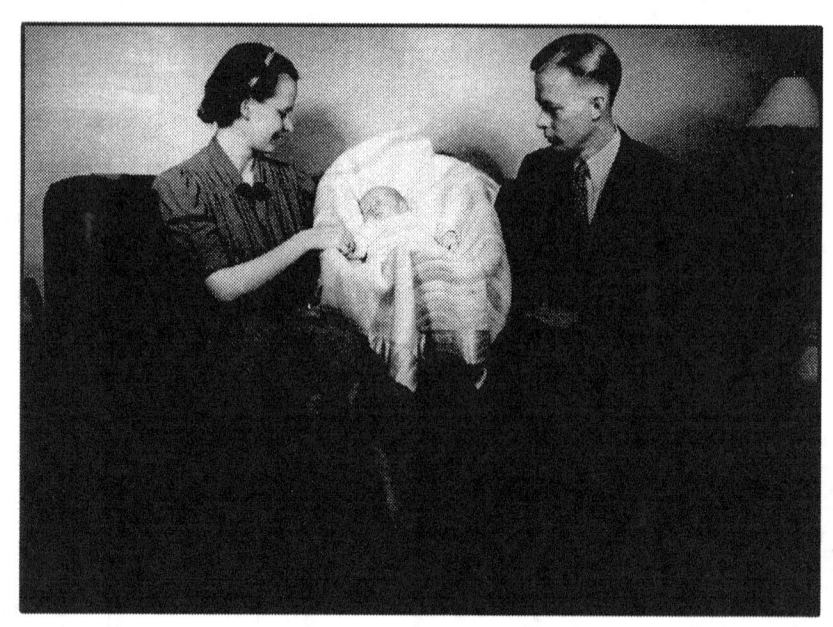

September 1940

Grandma Gertrud and Alma
(left)   (Right)

Berkley, father going to work on WWII Liberty Ships

Dad ruined his shoes saving me from drowning

Mom ... She grew up above Smith's store

Band Leader

Roof blew off the barn the night my dad was born (1909)

My kids, Kim and Eric exploring original homestead

My dad's first job, as a butter maker

**Private First Class
John Strand**

Strand Market

Inside the store

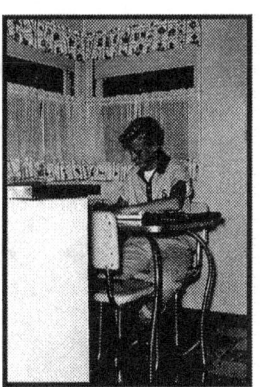

**Tract homes soon replaced
these orange groves**

**Algebra Homework
behind market**

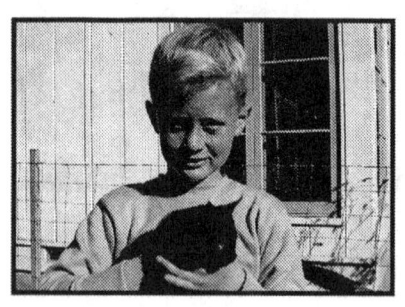

Tijuana Tuck and Roll

First Car

CHAPTER 1

THE EARLY YEARS

I was born eight miles from the Canadian border in Hallock, Minnesota. That northwest corner of Minnesota was known for its harsh winter weather. In 1942 my parents moved to Berkeley, California, where my father worked on World War II Liberty ships. Then, when I was seven, we moved to Baldwin Park in southern California.

Baldwin Park was named after the flamboyant land developer Lucky Baldwin ("Lucky," because of his success in mining). The main house of his 50,000-acre ranch was located where the Los Angeles County Arboretum and Santa Anita racetrack stand today.

Baldwin brought to the area irrigation watr; he then planted a wide variety of fruits, nuts, avocados and oranges. Pepper trees were started in many locations because of his efforts. Those very large pepper trees were perfect for a seven-year-old to climb and swing on. Orange throwing fights were a daily occurrence as kids explored the orchards. During the ensuing years, as I grew up, I saw the orange groves bulldozed to make way for tract houses. Baldwin Park was growing in the post war boom. I really missed the groves, but not the smudge pots that the ranchers used to keep the trees warm on frosty nights. They were burning reclaimed oil and old tires. It was not unusual to wake up in the morning with black soot in our noses.

I had many childhood friends and we went from house to house during the day. The fever that the first neighborhood television sets created among all of us kids could only be described as near pandemonium. At night we all gathered outside the windows of the three homes on our block that had televisions. Sometimes we were lucky and got invited inside to sit on the rug and watch. We would fight outside the windows for the best view of an eight-inch black and white television. "Time for Beanie," a puppet show, or especially "Hopalong Cassidy" would create scuffles outside the windows. The blinds were pulled down more than once. We only climbed higher on the fence to strain for a peek between the cracks of the blinds.

I nearly wore out my welcome at a friend's house where I showed up a couple of times a week. We watched while the stars circled the test pattern tower when the regular programming started at 5 pm. I recall overhearing my friend's mother saying to her son: "don't let any other kids start showing up here like John." I worried about this, but I was never asked to leave.

Marbles was a spirited game we played at Margaret Heath Elementary School in Baldwin Park. My second grade quality recess and lunch was usually spent around a circle drawn in the dirt. The "pot" as it was called varied in diameter from two to three feet. Anyone with an ante (usually two or three marbles) could play. You lagged a marble to a line four or five feet away to see who went first. You could also build a small mound called a "peak" under your target marble. You curled a little loose dirt up with your little finger then tamped it down with your palm and set the marble on top. The idea was to shoot a marble by flicking it off your index finger with your thumb. You sort of dug your thumbnail into your finger until you had built up some elastic potential energy.

The marble, when triggered, would launch into the air toward one of the marbles in the pot. If it struck a marble solidly and the struck marble was ejected from the circle, you got to keep it and then took another shot from the location where your "shooter" came to rest. Your shooter was usually your favorite and most effective marble and could also be referred to as a good "sticker" for its tendency to remain in the pot. The most impressive shot was an air shot with no bounces and a crisp clear crack when the targeted marble was struck.

It was roughly equivalent to a pool game where English and position strategy are important. Occasionally you could run the pot on a single turn. You got to keep every marble that left the pot. But to keep shooting you must get at least one marble out on each shot. Each player took a turn from outside the circle till all the marbles were gone. The last player to hit the last marble out of the pot was vulnerable to the "pots rule." If your sticker stayed within the circle while knocking the last marble out another player could yell "pots!" He then had the right to pick up your favorite shooter. Of course, if you yelled "no pots" first, you could keep your marble. Your shooter often represented a statement about yourself and was always carried in a pocket separated from your regular marbles. Some were cracked or scratched from long use, but when they spun to a stop inside the pot they were gorgeous. Many favorite personal shooters wouldn't be risked on the last marble in the pot.

Some marbles seemed to have a special intangible quality for sticking in a pot. They were usually slightly larger in size; not "boulders" as marbles nearly an inch across were called, and certainly not "pee wees" as very small marbles were called.

"Dubs" or "no dubs" could be yelled out if two or more marbles rolled out on a single shot. If "dubs" were shouted the extra marbles would be returned to the pot. But if the player managed to exclaim "no dubs" first, then he got to keep all the marbles outside the circle. Also if a player crossed the line during his shot or tried to move his marble position during his turn, "fudging" was called and the player would lose his turn.

The color and design of marbles was also thought to carry some extra special property connected to success at marbles. "Purees" were totally clear in a variety of colors and were valued as possessions, but weren't thought to be all that great as stickers since you couldn't see them spin. "Cat eyes" and "cat ears" were very beautiful with multiple colors and intricate interior structures. "Black beauties" were like black pearls, "rubies" were valued like the gemstone and "agates" resembled those stones found on certain beaches. "Steelies" (ball bearings) weren't too useful as they were too heavy to shoot and were sometimes banned from games. There were many beautiful, beautiful marbles without names and some which may have escaped from Chinese checker games. It is the strife over the marble game that stands out in my mind. Disputes leading to fights were a common occurrence.

One day at school something totally unexpected started. Someone dug a small hole in the ground and offered to give you several marbles if you could lag yours in. The operator kept marbles that missed the hole. It was a new type of marble game that caught the skilled players off guard. No longer would they dominate the marble games and it seemed to give everyone a better chance. To understand how this happened you must consider the world of the chronic marble loser. Virtually all games had the "keepsies" rule (loser's weepers). "Funsies" might be played in your neighborhood where mothers ruled, but the school yard game was strictly serious. The marble player, like the golfer who buys and doesn't fish for his golf balls, must face reality. Bring your purchased marbles to school and, unless you are fairly skillful, you are probably going to lose them.

In a way, marbles represented cash or trade value to us second graders and there was usually an incredible commotion connected to the fortunes of the game. Thus, when the novel new game was emerging, traditional marbles would never be the same. It started as a harmless sort of event involving a hole in the ground and a prize. Standing behind a line you would toss your marble toward the hole. If your marble made it in you got two, three, or four marbles in return. The distance from the hole and the size of the hole determined the payback. Clearly this was Las Vegas style marbles, and no one would set this up unless they had the odds on their side.

This game, which was operated like a carnival stand, began to get everyone's attention. Soon there were dozens of these little enterprises and competition was fierce to attract players. Some tried offering more and better marbles, larger holes but with obstacles in front. No one was playing hopscotch or kick ball or even ordinary marbles. There was more than the traditional frenzy to get out to the playground at recess in order to establish the best locations. Running and pushing in the halls was almost impossible to stop, as the teachers were to learn.

I had been able to amass a substantial hoard of marbles at home, but found they were gradually migrating back to some of the less skilled players who operated the "hole toss."

By the time it occurred to me to open up one of these marble making businesses, a new and even more spectacular innovation of the game developed.

Pitching pennies to a line (closest wins) had been around for a while. It interested some of us occasionally. Then a seven-year-old entrepreneur started offering pennies instead of marbles as the prize. Pay a penny, roll a marble and win cash. Marbles became an unimportant facilitating medium for the new sport and those offering marble prizes lost customers to the real gambling.

To understand what happened next you need to understand the role of the principal in the life of a second grader. He was to be avoided at all cost. I never used the hall that went by his office. I preferred to go outside in the rain to the other end of the building if necessary.

I came to understand the enticements of gambling at a very young age. I don't know if all this detracted from our academic progress. I do know that the recess and lunch teachers couldn't get students together to participate in anything else, all we wanted to do was to roll marbles for pennies.

There had always been problems with traditional marbles... poor losers, arguing over rules, cheating etc. Tears weren't unusual as the skilled players looked for easy victims to fatten up their pockets. Teachers could step in and settle disputes occasionally, but generally we preferred schoolyard "might makes right" justice.

Money was the inflammatory ingredient that would eventually destroy our entire marble economy. I imagine it began when some children went home hungry because they lost their lunch money down one of the holes. Parents learned of this; ergo: teachers learned of this; ergo: the principal learned of this; ergo: the principal who was seldom seen walking around our play yard was spotted kicking dirt into our little holes. And that ended that.

I don't remember if regular marbles were actually banned, but all our holes were filled and no one seemed interested in marbles any more. We were all sort of listless at recess after that. We slowly walked down the halls once more. Most of us just sat around and waited for lunchtime to end.

My parents purchased a small electrical shop one block from the Santa Fe Dam. They turned it into a grocery store and we lived in back. The wholesale grocery house was located in Pomona. Car production had been stalled during the war, so cars were hard to come by. My folks owned a 1936 Dodge that usually had to be pushed before starting. The ride over Kellogg Hill to Pomona regularly put that car to the test.

Down the street the rumble of rocks being dumped to complete the Santa Fe dam was heard every day, and even though Aero Jet was ten miles away, we still noticed the frequent sound of rocket motors being tested. Occasionally the rocket sound was cut short and a large explosion was heard. Nearby were the rock crushers that exposed massive holes in the ground. They were well over a thousand feet long and hundreds of feet deep. This is where the sand and gravel came from for all the new homes being built throughout the San Gabriel valley. Some of these pits remain to this day. Some were filled with trash and one was seriously considered as a location for a Los Angeles Raiders stadium. It was a tempting area for us boys as we would slip in at night to climb the crusher tower and catch pigeons.

The local residents in the late 1940s used the desert chaparral near the crusher pits as a dumping ground. There was trash and old appliances strewn as far as the eye could see. Attitudes about the environment were far different then. I remember Seal Beach in Orange County where there was a section referred to as "Tin Can" beach because you could hardly see the sand for all the empty cans.

Since there were no litter laws in effect most people perfunctorily discarded cans, bottles or paper from their cars. As kids we would walk along the roadsides and collect the soda bottles. A wagon full could net you some real cash. We got two cents for a small bottle and five cents for the quart size bottle. Not bad for a day's walk and you could still buy a small coke in those days for five cents.

Telephone service began offering long distance (you could have any color phone as long as it was black). I well remember a very emotional and tearful conversation between my mother and her twin sister in Minnesota. Even though they had kept in touch through letters, they had not talked with each other in five years.

Saturdays everyone packed the theaters to watch cartoons and a Roy Rogers movie. Flash Gordon and Buck Rogers' serials had you returning the next week for sure. All that for just twenty-five cents!

Even the shoe stores were changing. There was still only one type of tennis shoe, but now they had x-ray machines that showed the bones in your feet. What fun it was to be able to see your feet through your shoes. The salesman would also put his feet in the same picture to demonstrate a correct fit. It was an eerie green picture and I found it exciting to watch as I wiggled the bones in my toes. This little machine of course didn't last long as the salesmen and customers were getting high doses of very powerful radiation with its cumulative side effects.

Returning veterans and mid-westerners fleeing colder climates could purchase a three-bedroom home for five thousand dollars with only one hundred dollars down. I remember an orchard owner donning a clown outfit. He stood out on Garvey Blvd. waving a sign trying to snag anyone to purchase one of his lots for a thousand dollars. The San Bernardino freeway and a major mall are now located in close proximity to those orange grove lots.

My uncle, Al Nordling, had returned in 1945 from a German prison camp. The Russians had freed him as their troops entered Germany. He was shot down in a B-17 on the way to Ploesti. Anxious to get home he rode a PT boat from England to America. His severance pay allowed him to buy an orange juice stand on Garvey Blvd. in Baldwin Park. This was where uncle Al met my aunt Mame. While vacationing from Ohio, Mame and her brother had car trouble across the street from the juice stand. They met and were soon married. The juice, olive and date stand was remodeled into Nordling's market. Next door, a drive-in hamburger stand opened. Aunt Mame sold them the ground chuck and potatoes for their first hamburgers and fries. She also contributed a key step in the hamburger recipe. This became the very first IN AND OUT BURGER, now in hundreds of locations and is historically considered to be the world's first drive-in restaurant. On their first logo tee shirt is a picture of the original burger stand and in the background you see Nordling market. Needless to say, I ate one of the first hamburgers served on opening day of IN AND OUT BURGER. Wow!

Local dairies were a part of the economy. Barbara Littlejohn from my high school class was raised on a dairy near the western edge of Baldwin Park. She married Dallas Long, a UCLA graduate who held the world's record in the shot put. He received a Gold Medal at the Olympics. Kent McCord, officer Jim Reed in the Adam-12 TV series, was a grade behind me. General Jimmy Doolittle had family living down the street from us and was believed to have visited our neighborhood. A few other celebrities grew up in Baldwin Park. The James Bond movie actor Richard Kiel (Jaws) was a friend of mine in high school. Steve Gibbs, owner of the National Hot Rod Association was also a close friend.

High school was a nightmare. It started in the freshman year with a tradition called "smearing." Just as we were learning to cope with the routine of changing classrooms every hour, we freshmen had to watch out for the upper classmen. One or two of them would hold you down while some bright indelible shade of lipstick was smeared on your face. The teachers did their best to prevent this on campus. It became a cat and mouse game, trying to walk the hallways only when teachers were present. If you made it through the day the next hurdle was leaving school. The seniors were waiting for you and there was usually no safe way to get home. In fact, most of the kids walking home would usually stop when accosted and submit rather than be chased down on foot. The school buses were where the real action happened. Carloads of taunting students would buzz the school bus rapping up their twin pipes and shouting threats. When the bus stopped it resulted in a shriek from the girls and pell-mell foot races to reach anywhere safe. My plan had worked fairly well for me. I would wait until the last student left the bus and then streak away as the tormentors were chasing the others. My luck ran out one afternoon while on my newspaper route. A senior came out of one of the houses and told me to wait right there while he returned to the house for his sister's lipstick. I regretfully waited and `was "smeared."

Baldwin Park High was a seemingly ordinary institution. The only issue we ever protested was when the school banned Levis. We marched around the campus at lunchtime and the administration canceled the ban. Flattops and ducktails were our hairstyle. We wore leather jackets, tee shirts with the sleeves rolled up and Levis or a pair of peggers. The girls liked the men's Levis too. They wore them on Saturdays with their dad's best white long sleeved shirt. All the girls went shopping on Saturdays with their hair up in rollers. One can only guess that this was the way to tell everyone they had a date for Saturday night. And yes, you could do anything, but, "don't step on my blue suede shoes!"

A Model A Ford full of gas sold for fifty dollars and would be "souped-up" with a larger overhead valve engine. These were our street legal Hot Rods. The engine and parts were probably found at a junkyard near five points in El Monte (convolution of five roads). Regular sedans had the chrome ornaments removed (nosed and decked), plus a tuck and roll interior installed in Tijuana, Mexico.

Words like cool, cat, square, hip and groovy developed new meaning. The action was at Bob's Big Boy drive-in located on the Rose Parade route in Pasadena. It was worth waiting a half-hour to cruise through in your "hot" car, showing off your candy-apple simonized wax job. Flames and scallops with pinstripe painting started in Downey. Lowering a car in the back and skirts over the rear tires was out. Raking or lowering in front was in. Twin glass pack mufflers were a must. Drag racing was being organized at strips out in Pomona and in Irwindale on the eastern edge of Baldwin Park. Street dragging and the excitement of hubcap stealing were experienced by most of the "in" crowd at least once. There was a small hubcap black market at most high school parking lots.

Beer was the main course for our parties that were sometimes called beer busts. A favorite jumping dance spot was the El Monte Legion Stadium. There you might rock out to Johnny Otis. The fox trot was our slow dance. Jitterbug was out as we invented our own style of fast dances. The "bop" was in. Elvis, along with a list of other rock stars, was dominating the exciting new pop music. "Huggy Boy" was our late at night disc jockey. He played all of the new rock and roll tunes.

Even though the dances were fun, a serious date usually meant an evening at the local drive-in theater. Later a cruise up to Chantry Flats above Sierra Madre to view the city lights. Other destinations were Disneyland or Santa Monica's Pacific Ocean Park. And who could resist going for a ride on the "Rocket" roller coaster at San Pedro?

I was lucky enough to have saved money working as a box boy and mowing lawns. I bought my first car (a 1952 "Merc") and of course got my first ticket shortly after that. Seems I was doing sixty-five in a twenty-five zone. I don't remember who won the drag race, but I do remember we both got a ticket. My father never got a ticket in his entire life, so imagine my embarrassment when the judge suspended my license for six months. I was also informed after the six months was up, that if any officer in the department spotted me driving down the street, they were to pull me over. He was not kidding. From then on in Baldwin Park I was pulled over once or twice a week and let go. My custom car with its lacquered paint job was very easy to spot.

There were regular, almost daily, fistfights at school. If it occurred at lunch, nearby teachers usually stopped it. The gym teacher, Mr. Snell, had an interesting technique he used to stop a fight. He would come up and watch for a while. When it appeared that someone was winning and about to hurt the other person seriously he would kick the winner hard in the rear. This would stop the fight and he would collar both of them, taking them to the office.

"Meet me after school in the parking lot" were the worst fights (choose off's we called them). During school hours the threats were usually made, then for the rest of the day sequestered groups talked about the two combatants. What would happen? What were they going to do to each other? What was the record of each of them in previous fights? These discussions went on all day. Anxiously everyone waited for school to let out and then dashed off to the school parking lot to watch the fight. On Fridays there could be two or three fights at the same time. The crowd often couldn't decide which one to watch. Teachers would finally arrive and bruises could be counted. Fear of a confrontation with other students reached a zenith by my senior year. I remember turning around to a full right-handed "haymaker." What made it even worse was an acquaintance was used to catch my attention while the attacker got behind me. Changing classes was always a tense time. I had to dash down the hall watching my back at all times. I would avoid going to school whenever possible. My senior year I remember getting six swats from the Vice-Principal because I earned six demerits from skipping school. In order to graduate I had to bend over and grab my socks while a large paddle was applied.

An incident that happened over a bottle of wine at my folk's market was about the only crime I remember. The local drunk was apparently not able to amass enough money to buy his daily bottle. I usually saw him sleeping in the weeds at noon or staggering around begging for pennies. One day he walked in and grabbed a bottle of Muscatel wine off of the shelf. As he ran out without paying, a surprising response overtook my petite mother. She took off after him and a tug of war over the bottle ensued along the busy street. She finally kicked him between the legs and flattened him. Triumphantly she returned the bottle to the shelf. Only then when the adrenaline wore off did she realized what she had done.

When I graduated in 1957, I had shown no aptitude whatsoever for mathematics. In fact, the only "D" grade I received was in intermediate algebra. This bothered my mother, who had always taken a genuine interest in my schooling. When I got a bad grade she always talked with the teacher about it. When I fell behind in reading she got a "phonics" book and taught me to read. But math was something else, she never succeeded in helping me very much here. I remember sitting at a table for what seemed like hours with an unsolvable word problem. When my tears came, as they frequently did, I could go to bed.

I always thought it funny that back in Berkeley, when I was in kindergarten my teacher seemed enthused about my abilities. She made me the conductor for our fifty-kid orchestra. I wasn't sure why, certainly not for my non-existent musical talent. Of course, as conductor I didn't have to play any musical instruments.

I was disinterested in anything relating to high school. You can't find me in the yearbooks, as I refused to show up for pictures. I never bought a yearbook, did not join any clubs, didn't go to the prom, and no class ring. I could not wait for the whole mess to get over with. I certainly had no intention of going through the graduation ceremonies. The counselor contacted my mother. "Mrs. Strand, do you realize that out of a class of three hundred and fifty graduating seniors only one young man has refused to buy a robe?" Did she straighten me out on that one? You bet cha!

I was a box boy for Shopping Bag in La Puente. That's where I met Harvey Pelletier. Among all of my friends, Harvey was the first to get married. He had heard that a ship loaded with French tourists was docked in San Pedro. If you went down to the docks you could talk to the tourists through a chain link fence. Harvey hurried down to the docks and immediately struck up a conversation with one of the pretty young girls. He convinced her that he wanted to marry her right then and there. With the help of the captain of the ship, Harvey went home with a new French bride. She bore a striking resemblance to Sophia Loren. This really took us all by surprise. Back then we were still having trouble getting up enough nerve to ask a girl out and here came Harvey with a bride!

One of the remarkable things about Harvey was his large and powerful buckteeth. He could pull the top off a bottle of beer in a heartbeat. Not with an opener, but with his teeth. I had seen this performed once by a person who used his molars to slowly loosen the cap. Harvey did it straight on with his front teeth and with one quick snap.

This came in handy at parties where I remember a line of people in front of Harvey. While Little Richard sang "Tutti Fruiti" Harvey was popping the bottle caps off.

Looking back we were establishing the story portrayed in the movie "American Graffiti." Nostalgia is very strong in me. I dream that someday I may return to Baldwin Park and see all the tract homes gone. I would then find that neat little rows of orange and avocado trees were flowering. Irrigation trenches are full of water and swarms of sparrows are singing. I chase a jackrabbit into cactus briars near a giant yucca flower. I'm careful not to step on the red anthills as I see a horny toad scurry away. I start to climb the dam and wave to my friends Hitchcock, Fetterolf and Henson from my Boy Scout Apache Patrol. Rain has created a small lake behind the dam with ducks swimming there. We spot a place we have never explored and if we are lucky it might even have a tiny cave to crawl through. Life seems very simple and wonderful.

After graduating I began to get restless. Minor disagreements with my parents were increasingly more common. At seventeen, shy and naive, I was ready to enlist in the Army. My mother agreed to sign the paper.

Basic training was at Fort Ord, California. The Army turned out to be no better than high school. Everyone got into at least one fight. My slender frame started filling out. We got three full meals a day, plus workouts, long marches, tear gas, hand grenades, machine guns on infiltration courses. At breakfast I would return for six more eggs. I gained twenty-five pounds in ten weeks and it was all muscle. I did the maximum that the Army hoped for, sixteen pull-ups, one hundred squat jumps and sixty pushups. I was physically filling out at "warp speed."

A new self-confidence was emerging. I was at least winning some of the fights. Not that I looked for fights, but the Army brought together the most amazing diversity of people. I defeated the company bully in a dark vacant lot beside the barracks. I had speared him like Dick Butkus tackled. His friends told me I was lucky because this guy knew "jujitsu." My fellow soldiers cheered for me and partied while I cried. I felt bad about the fact that I had hurt him. His shoulder was separated and he was headed for the hospital. Then I met Ambrose, a semi-pro boxer from Oakland, California. We became good friends and hung out together. No one bothered either of us after that.

Finishing basic, I was transferred to Fort Sam Houston, San Antonio, Texas. My duties included attending the hospital and battlefield medic school. Next I was sent to the Medical Laboratory Technician school where I first exhibited an aptitude for science. I got the top score among two hundred fifty lab technicians in the sixteen-week course. I loved doing the white cell analysis, especially cell differentiation. With glass templates and thin glass cover sheets we counted white cells. The Van Slyke carbon dioxide test was the most complicated. These methods are all archaic today. We each got to check our own blood, and I had the highest Eosinofils among my classmates. I always excel! Our instructor thought that my count was high because I was raised in southern California. Going barefoot for most of my childhood meant I likely picked up some harmless parasite that caused my Eosinofils to increase.

There was one night exercise that I found strange. Our company commander had either goofed or was purposely subjecting us to intense training. I never found out which. It developed this way. The days were warm enough that you could skip wearing a field jacket at morning formation. All of us formed up one morning and without warning busses arrived to take us to a field exercise. I was a casualty for a medical evacuation. A chopper airlifted some of us and the others were cared for on the ground. C-rations were handed out for lunch with no can openers. This should have given us a hint of what we were in for. We waited several hours for evacuation, but no busses showed up. Lunchtime passed and we had lots of canned food but still no way to open them. As we got hungry and eventually desperate we found that bayonets were not really good instruments for opening cans. The only method that seemed to work was to smash the cans with rocks. After removing the dust and dirt from the top layers we could eat what was left in the can.

We were hungry again as night approached and no more food. As the weather got colder we realized that most of us had not brought our field jackets. We had been given no warning about how long the exercise would last. Scuttlebutt was that we might be getting a lesson in preparedness, what with no warning that it was going to be a night exercise. We were jumping up and down to keep our bodies warm. The sun disappeared and it got really cold as is usual for Texas in December. Finally we were transported by small vehicle to an old

PX many miles from our barracks. The PX was closed so all we could do was gather around outside. With no one in charge we were 250 hungry and cold privates. The last jeep driver told us that the earliest you could wake up a bus driver was 5 am. So it looked like we were on our own for the rest of the night.

As the temperature dropped below freezing, jumping around no longer kept us warm as numbness set in. A fire was lit beside the PX. Gradually anything that would burn was thrown on. A nearby fence was ripped apart. With a large group and only one fire, the inner circle roasts while the outer group freezes. One at a time we would dash in close to the fire, spin around and then run off with our clothes steaming. This got old very fast. Pieces of the PX were beginning to show up on top of the fire. Any loose board that could be yanked off or pried off with our bayonets was fair game. Thirty or forty-fire dancers remained around the fire while the main body of the company formed a wedge into the "L" of the building. It was slightly sheltered from the wind and two hundred men pushed against each other to keep warm. There were minor injuries to the men close to the building, but they were the warmest. Frostbite was possible, but as far as I know, did not occur. I still remember how warm the buses were when they picked us up at 5:30 the next morning.

I am still not sure if this type of harassment of new recruits was beyond Army regulations. The company commander did remind me of the wild-eyed, white-haired "Original Bad Bob" straight from the movie "Judge Roy Bean." You might remember the scene where Bad Bob ordered his horse to be cooked for lunch just before Paul Newman ambushed him in the back. Our company commander was replaced shortly after that overnight exercise. I also noticed that the Inspector General became a more frequent visitor to our barracks.

My thirst for knowledge was increasing. During my on-the-job training at Fitzsimmons Army Hospital in Denver I began to read Scientific American. I remember an article that fascinated me about the orbit perihelion of the planet Mercury. Its precession was violating Newton's law. Einstein predicted this. I had no idea that ten years later I would contribute to another verification of Einstein's Theory.

Learning to understand antigen-antibody relationships from the Army field manuals was an interesting exercise. On one of our field exercises a medical doctor told me that we had the only field laboratory he had ever seen that performed good laboratory tests. I spun the centrifuge all day with my arm. But the field exercises turned out to be a disaster for me. I got twenty-five major tick bites. These bites bothered me for decades. Not to mention waking up one morning to find a rather fat six-foot rattlesnake right outside our tent! The snake had been run over by a truck the night before. It had been thrashing there till we found it at first light.

Upon my release from the Army, I returned home, found a place to live and enrolled at Citrus Junior College (AA in Chemistry). Taking a much-needed break, I was visiting my parents. As I paced back and forth with mathematics on my mind I glanced out mom's kitchen window. "Baby in the pool!" I shouted as I raced out the door, planning to jump from the second story balcony. I quickly changed my mind and took the steps four at a time. Lounge chairs and umbrellas went flying as I leaped into the water with such fervor that I nearly knocked myself out on the other side of the pool. The baby, bobbing up and down in the water, now began to cry. I pushed her up and out of the pool just as my parents and the frantic mother appeared. After a big thank you the mother took her wet, but safe, two-year-old home.

Returning upstairs my dad began reminiscing. When I was about two years old, dad had to jump into a Minnesota lake to save me from drowning. It cost him a pair of brand new oxfords. World War II rationing had made those shoes hard to come by!

As a chemistry major I renewed my youthful interest in explosives. In class we used gunpowder as an ingredient for experiments. The gunpowder was kept in a large crucible and we were able to help ourselves as needed. Some fellow students were playing with a glass rod and a Bunsen burner. They would touch the glass rod to the gunpowder and then go to the flame on the burner. It would "snap, crackle and pop." Back to the gunpowder, then back to the burner. This went on and on, back and forth, until one young man returned the glass rod too soon to the gunpowder. The tip was way too hot as he plunged it into the gunpowder. The large crucible of gunpowder detonated. The students seemed to come through it OK, except for some singed eyebrows. The ceiling remained black for years. After that, the instructor merely pointed to the ceiling when explaining why gunpowder was no longer allowed in class.

My roommate, who was also interested in things that go "bang," showed me the process of making nitro glycerin. It was surprisingly easy. We began setting off little explosions in our back yard. One afternoon our neighbor looked over the fence and casually asked what we were doing. We told him about the small explosives we were making. He excused himself and said something about going to the store. We set off a few more and then decided to increase the ingredients. The last one contained almost half a shot of nitro glycerin (I actually used shot glasses) with cherry bombs for detonators. I had to light the fuse with a match and run very fast. My roommate was in the house under the bed with fingers in his ears. When it detonated I could feel the concussion behind me as I was running away. Then I heard it: "the tinkle of glass." I had blown out the neighbor's bathroom window. His only comment when he returned was:

"It's OK John, I was going to remodel the bathroom, anyway."

The leather jacket I almost lost

I took Steve out to the drags for his first race

CHAPTER 2

WALKING SKYSCRAPER

My first recollection of Richard Kiel was at our grocery store on Halloween. He had made a cardboard extension to put on his shoulders. It rose two feet above him with an ugly cardboard face atop that. Richard was on his way to well over seven feet tall before he put on this costume! The extension on his shoulders made him a walking skyscraper. He was wrapped in a dark heavy blanket dragging a large chain from his leg as if he had escaped from who knows where. I still get chills thinking about it. His immense height and that chain dragging slowly behind him indicated he was not your average "trick or treat'r." Children fled in terror and disturbed parents got out of his way. There were a couple of brave twelve-year-olds that dared to get within ten feet of Richard, but a heavy groan from him sent them running off.

Richard and I both grew up in Baldwin Park. His parents owned Kiel's Appliance in the downtown area. My parents owned a "mom and pop" grocery store. Richard would become a major movie star and one of the most well liked people in movies. He was the only villain to be in two James Bond movies. Playing the character "Jaws," he was given steel teeth and could survive a fall from an airplane.

Richard played Junior Varsity Basketball at Baldwin Park High, but was not inclined to take up the sport seriously. We did work together at his folk's appliance store. We were a team. I opened the doors while Richard picked up the washer, dryer or refrigerator and carried it in.

It was around 1956 that drag racing became a spectator sport. Richard Kiel, Steve Gibbs and I all were very serious about our cars. Steve also attended Baldwin Park High. He got his first good job as the Drag News reporter and also worked as a starter at the Irwindale raceway. There were a few amusing incidents involving Richard and his 1951 Ford. He had to move the front seat back so far that it pushed against the back seat. This was the only way he could fit in the driver's seat and even then it was with difficulty.

There were of course, rumbles and car posturing back then. Most of it was yelling and gesturing. Never aggressive, but when Richard was challenged he would pull his car over. No matter how many kids were in the other car, when he uncoiled slowly and stood beside his Ford, they would drive away quickly.

One of the most memorable adventures with Richard started when we decided to exchange coupons for money at Market Basket. Discount coupons, which reduced the price of canned goods, were a regular distraction to checkers. They took time to count and usually paid the store a penny or two for handling. It was such an annoyance that checkout clerks began to accept any coupon and quickly give you face value. They quit looking to see if you purchased the item. Eventually they didn't care if you bought anything; they just cashed you out. This sets the stage for the great coupon panic of Baldwin Park. The magazines Life, Look and Ladies Home Journal were carrying Swift Premium coupons one month. Swift Premium was not aware of the developments with the disgruntled Baldwin Park merchants when they printed a page of coupons worth a dollar forty-nine in a twenty-cent magazine. I heard about it from Steve who worked for the market. Wow, over a dollar profit on each magazine we purchased! At that time I had to work an hour as a box boy to make a dollar. It appeared to be legal, as the small print on these coupons did not forbid it. The excitement that ensued was to grip Baldwin Park for the next several days. Children with mouths and pockets full of candy, parents smiling with free groceries and stalled checkout lines of anxious coupon clippers. It was not uncommon to see a customer bring in an inch or two of coupons and walk out with fifty or sixty dollars. Then back to the store for more magazines … you get the picture.

Like a gold rush it spread. I felt like one of the last to hear about it and in a near frenzy I was able to generate a little seed money. Soon the magazine supply dried up in Baldwin Park. Steve had to work, so Richard and I discussed these developments the next day at school. Before I knew it we left school for more important "things." We did locate some magazines in a nearby town and realized this was not going to last very long. Instead of scissors we found razor blades were faster slashing out the coupons. A dollar forty-nine in less than a minute! No school for us the next day either. Panic was setting in and we had to drive further for magazines. Richard called a magazine distributor and he was asked "how many do you want?" Knees shaking, we called the distributor back in an hour. He said he wasn't sure what was going on but there were no magazines available in the entire Los Angeles basin. This meant we had to find pockets of stores where the news hadn't spread. Beverly Blvd. turned out to be rich in Ladies Home Journals. Supplies of Life and Look that were 15 cents cheaper had by now dried up. We purchased twenty to twenty five copies at a time. We told the puzzled clerks that it was our mother on the cover.

Richard's back seat began to fill up. We were starting to lose vision out the rear window. As it got later in the evening we were a long way from home. We found ourselves in downtown Los Angeles nearly out of gasoline. To save on gas Richard turned off the engine and we coasted into the underground Pershing Square parking garage. It was around midnight and here we were on "Skid Row." As he went off to find directions to the nearest gas station I headed for the restroom. It was very still and quiet with no one around. All of a sudden I realized two big guys had followed me in. One of them spoke casually to the other "Do you want a leather jacket?" I glanced their way and noticed that I was the only one wearing a leather jacket. Talk about timing! Richard came through the doorway momentarily. Actually he just stood in the doorway; no, he was the doorway! He just quietly said "John, let's get out of here." My leather jacket and I left with Richard.

We returned to Baldwin Park with the car full of magazines. Missed another day of school cutting out coupons. But what we found was that the stores were now requiring you to purchase something of equal value. Our stack of coupons was six or seven inches tall. New game plan. We purchased canned goods with our coupons and then searched out a small grocery store. My parents had sold their store by this time, but I could never have approached them with our scheme anyway. We found the new owner would purchase the canned goods from us at a reduced rate and there went our big profit margin!

Virtually all the magazines in the Los Angeles area were now gone. We considered checking out some other states, but decided to go back to high school. Our dreams of careers as coupon entrepreneurs were dashed. We did look the next week but there were no coupons in the magazines. To this day I do not clip coupons, I have no stomach for it. It would be weeks before a Journal could be found anywhere since it was a monthly magazine.

Resigned to working for a living, Steve became the owner of the NHRA, I became an astrophysicist and Richard became the movie star.

Didn't believe the palm trees were real

Palm Springs on the side of mountains

Lower mountains weren't as steep as Mount Hacinto on top

Thanks to the Palm Springs Visitor Center

CHAPTER 3

AS THE CROW FLIES

The merits of deer hunting are not the issue here. I have been a vegetarian for several decades. Many decades ago I did like to hunt and got myself into a bind the likes of which I would never experience again, thank God. My best friend Raymond Osgood was a hunter in every sense of the word. His father was a hunter and his father's father was a hunter.

Raymond did have a way of getting me into scrapes, but usually they were minor. Like the time he caught a large King snake and talked me into putting it into my car glove compartment for safekeeping. About five miles from home we checked the compartment and the snake was not in it. As we turned the corner onto the block where we lived, the snake dropped from under the dash onto the gas pedal and my foot! I do not like snakes, period! I jumped out of the car leaving Raymond and the snake to continue down the street. Raymond was in the passenger seat trying to steer and stop the car. As Raymond got control of the car the snake disappeared back into the dashboard. The tangle of wires and whatnot made it almost impossible to locate the snake. Several hours later Raymond's father, with a coat hanger, dislodged the snake from the defroster tunnel, grabbed its tail and wrenched it out of my car. I had already discussed signing the pink slip over.

Raymond had a friend in Hemet, California named Wilma. Wilma was an experienced Native American outdoorsman who arranged guided hunting trips. He owned a substantial ranch on the edge of the reservation. This wilderness went up the side of the mountain to Mount San Jacinto and down the other side to the outskirts of Palm Springs. This was a storied land with tall tales of gold nuggets lying out in the open. Sportsmen would shoot the deer and Wilma would pack them out on horses. He only charged five dollars a day for unguided hunting on his property. He built a small shack for overnight hunters on the steep slopes of this nearly impenetrable wilderness. If he had not built this shack it is possible I would not be here telling this story. Wilma was a character to say the least. He picked out a small tree each year for target practice. Every evening he would shoot at the trunk. After weeks of this the tree would finally fall over and Wilma would select another tree.

Through the years, a number of hunters had gotten lost and died; some had never been found. The only account of a survivor was when a small plane crashed in the middle of this ten thousand-foot descent. The pilot did manage to walk out to Palm Springs. No one lost from the Hemet side had ever survived the hike out--that is until Raymond asked me to go hunting.

Raymond and I planned an all day hunting trip on Wilma's ranch. Svende Jorgensen, a coworker at Avery Label, and George Beersley were invited along. Raymond wanted to start hunting at first light. This meant we had to leave El Monte early, as it was a two-hour drive to Hemet. Then there was the one-hour hike up the mountain before we could start hunting. Needless to say we were so excited that we only got a couple hours sleep before we left for

Hemet at 2 am. We arrived at the camp area in what felt like the middle of the night. With no thought we left our food in the car, grabbed our rifles and started up the hill.

Raymond almost had to carry George up the trail, as he wasn't much of an outdoorsman. Svende and I took the steep assent fairly well. We saw a large group of deer on top but no legal bucks. Later that afternoon the wind came up. It was strong enough that a startled coyote twenty feet away did not hear us. The weather turned foul and light snow began to fall. Svende and I decided to head toward the car for food. Raymond and George were on top of the ridge and had built a campfire intending to follow us shortly.

Remember that we arrived in the middle of the night and had hiked up several miles by flashlight. We had no general sense of which direction the car was. We intended to follow the trail back. The falling snow was changing the scenery and covering what was a vaguely marked path. The trail disappeared and we realized we were on an unfamiliar ridge. Thick fog blew in plus more snow, and from that point on we were lost. Raymond had not clearly explained to us that we were on the edge of one of the last unexplored wildernesses in California.

After strenuously walking back and forth we determined we were on a large extending ridge. It was very windy on top and we had to choose which side of the ridge to go down. This would turn out to be the wrong choice. Svende fired off four or five rounds from his handgun. It sounded like a cap pistol in the high wind. I fired my Label (French Army rifle) ... not much better, as no one was going to hear it.

Boy Scout survival suggests a last resort when lost: go down hill and you will come out somewhere. This is essentially correct, but it would take all the strength we had to get down.

Down we went into the fog. Somewhere below the wind lightened up and we were encouraged by the rapid progress we were making. Down, down, down... ever steeper then gradually the terrain changed. The pine trees on top were thinning out and brush was replacing them. Sometimes we had to back track and choose a better path down. Eventually we had to conclude that nothing looked remotely like the original slope we had ascended that morning. We should have long ago come out at Wilma's ranch. It was a hard choice; we had come so far down and to return would be hours of difficult climbing. Since we had no idea where we were, we decided to continue going down and hope for a break. Daylight was receding; hunger and thirst were increasing so all we could do was intensify our pace.

We recklessly ran, slid, jumped over bushes to get off the top of that mountain. The steepness continued to increase, as did the brush. We alternated between a brisk pace and stopping to pick our way around cliffs, always trying to find an easier route. Occasionally we would have to retrace our steps as the canyons got more impassable. Progress became more difficult as our choice of paths required sliding on our pants with no certainty that we could climb back up!

As the sun started to set we chose a very steep path which appeared the only way down. We studied the consequences and finally, passing our rifles down a slope we made a small jump that settled it. We were now fully committed to a downward retreat. We could no longer back out of this canyon. The return to the top was now impossible. As the last rays of sun were fading, Svende climbed a nearby bluff to get a last fix on where we were. I noticed a tree with strange berries the size of grapes. I saw animals had been eating them and from my Boy Scout training I knew they were probably safe. We had been twenty-four hours without food and I chowed down. The seed was the same size as the berry. I was chewing the skin that covered the seed. It was sweet but I had to go through several mouthfuls to get the smallest amount of sugar out of them. I was hungrily chewing when Svende returned and said there was a shack on the other side of the bluff. I gulped down another mouthful and followed him.

Sure enough, the metal roof and rough texture indicated a ten by ten-foot shed stuck out in this wilderness like a needle in a haystack. God's blessing... We were to find out later that this was a hunting cabin Wilma built for his hunting clients. Horses and mules descended by a path that we did not follow and were stabled at this cabin. The deer could then be packed out when the hunters returned to the top.

We eagerly approached the shack but it had a padlock. Western movies had strongly influenced me, so I thought: "I'll shoot it off!" Svende stopped me. Noticing a loose screw he pulled it out of the hinge and the door swung open. The inside was very simple, only bunk beds and a small wood stove. There were several tins of food. I remember eating Cling peaches from a rusty can; they were the most delicious peaches I had ever eaten. A can of tuna and a can of tomato juice rounded out our meal. We were able to pour tomato juice into Svende's canteen. We located a flashlight and some matches. I suggested we stay at the cabin but Svende had an important date the following evening. It still hadn't struck us as to the predicament we were in. Darkness was now complete; the full moon was out so we decided to continue on. As the clouds cleared we saw before us the next leg of the journey.

We were high above a desert floor. There was what appeared to be a patchwork group of small plains leading us eventually down to a flat desert. All the canyons dividing the flat areas were the problem. At a distance it looked far easier than it turned out to be. There was one glimmer of hope. In the distant horizon were small specks of light. They were on the far side of the desert but they did indicate civilization and they were moving. Svende and I discussed this and unanimously concluded these were car lights. The desert can be very deceptive especially at night. These lights were possibly thirty miles away. We had no idea where we were but we felt if we could climb down far enough we could stop a car and be rescued.

With the help of the full moon and instinct, we pushed on. We actually crawled over large clumps of vines. My feet were off the ground and I was swimming in vines. Was I scared? Surprisingly not, perhaps because of the tremendous exertion. The moving lights provided some hope. Two years later the Palm Springs Tram was built a few miles north of us. I revisited this mountain in 1997 and could not believe I came out alive. The caveman of Palm Springs, Doug Batchelor, (now an Adventist preacher) lived in the lower regions during his dropout hippie days.

Fortunately we had eaten something at the cabin because the brambles took their due and tore at us. We lost the flashlight and the canteen full of tomato juice. We ate the last scrap of food from the cabin and continued our steep descent.

We did not know where we were headed. The cold and rain mixed with snowflakes caused us to stop and build a fire with matches we took from the shack. We piled it high with dead brush and tried to get shelter from the wind. I crawled under some overhanging rocks but they were cold. If you stood beside the fire the rain could reach you. We were beginning to get desperate when the rain almost stopped and the moon came out again. It sparkled on the rocks and we were sure someone would see our fire. Not realizing how far away help was, the sparks that flew down the cliffs signaled hope to us. We waited wet and confused.

If we could see car headlights then they surely could see our fire. Exhaustion began to hamper our judgment. There were horses and search parties coming up the draw to rescue us. We yelled "over here" but it was just the moon playing tricks on the rocks. No one was coming, no one knew we were there. Raymond had to call my folks and tell them I was lost. My father met up with Raymond's father and they left for Hemet. Raymond and George had to sleep in George's car. George, anxious to leave, was tired of the outdoors and was worried about scratches on his new Oldsmobile. Raymond was thinking about where to send a search party when it arrived.

Cold and stiff we decided to move on, leaving a campfire and a sea of sparks behind us. If only a ranger would come up and ticket us. There were small pools of water on top of rocks left by the rain. Chasing away the salamanders, we could get a sip or two before the sediment clouded the water. The only game we saw was quail and I learned a fundamental equation of the outdoors. One .45 bullet + one quail = one pile of feathers.

Hallucinations started, as a cabin on closer inspection became a rock. Svende actually took a shot at it, and then came forth with a huge curse directed at the mountains. This concerned me a great deal as he started to get a very wild look in his eyes. This was wasting our strength, so after that we had to ignore anything that looked man-made. We were both starting to lose it; flat desert areas seen at a distance became more rocks and brush. Our focus was still to go down ever down. Help was certainly out there somewhere.

The lava rocks started looking like beds. I could see headboards and pillows. Svende said he was going to pick out a particularly comfortable bed (rock). I collapsed on the nearest flat surface, leaned back and closed my eyes. Suddenly the sun came up and a new day started. I looked around dreamily and discovered that I was laying on the edge of a sheer cliff. There did appear to be real palm trees nearby. We had seen them in the dark of night, but thought they were hallucinations. We stumbled ahead and found a pool of water that was near a real cabin. Civilization, hooray! We were in a small resort called Palm Canyon. I knocked on the cabin door and a pajama-clad man gave us several glasses of water. Now for the $64 question. "Where are we?" "Palm Springs" he said looking doubtfully at our rifles. I asked him to call the police, which he quickly did. Food was the next thing on our minds, so we headed for town.

I have never experienced anything like the next half-hour. At the time I was twenty-two years old and didn't know what the saying "deja vu" meant. The sun was glaring down on us, my face felt like the skin was stretched tightly around my head. The hike to town felt rehearsed like we had rewound a tape and were playing it over and over. I pointed to pebbles on the road and nodded as Svende agreed we had seen them before. All this ended when we spotted a restaurant. I ordered the first thing I saw on the menu and went to call my mother. The waitress let us put our rifles under the table. I ate my meal and ordered the same thing again. Then with our rifles we went outside to wait for my father to pick us up. Immediately we fell asleep on the parkway along the sidewalk.

My father arrived, the search team had been sent home and everyone said Palm Springs was the last place they would have looked for us. Since they were searching for us on the other side, my father had to drive more than fifty miles around the mountain to pick us up. We didn't walk that far because we traveled the way the crow flies. I have very little advice about how to avoid getting lost. Except to say it never entered my mind that it could happen to me.

This story would not be complete without telling you what happened to Raymond a few years later. It does illustrate how different circumstances create the same problem. I, on the one hand, remember getting lost with sixty boy scouts and we spent an extra night in the woods. So don't think it can't happen to you!

Raymond's problem, on the other hand, was very different from mine. He was deer hunting in Utah on a relatively flat terrain. There were pine trees with some brush and snow covered the ground. After hunting all day he headed out of camp an hour before sunset with just a light jacket. Flat ground with trees meant limited visibility and the light snowfall was covering the path. He was having difficulty getting back to camp, when he saw some fresh footprints. The light snow had not fully covered them, but then he realized they were his own tracks!

This was a classic example of getting lost. True to form when without any distant landscape for direction, a lost person will often walk in a circle. Raymond is a close friend and the most knowledgeable hunter I have ever known. It took him only a moment to quickly size up the situation and determine he would be spending the night in these woods.

I guess it is usually the case that someone who gets lost does not really plan for it. With only a light jacket, warmth was going to be a major problem. The light was fading and he needed to build a fire. In his pocket were some matches, a few scraps of paper and several dollar bills, really not much with frostbite and even his life on the line. Wet dead wood will burn if you have a kindled fire. He needed a small fire and fast. He found a rotting snag from which he was able to kick away the surface and get some resinous wood. Often this wood smells like turpentine. He got under the boughs of a tree for shelter. Like the adventure movie he was living, there would be only one chance to get a fire started. Only five matches, the matchbook cover, a few scraps of paper and three dollars were between him and freezing,

Well, Raymond hit a home run with two outs in the ninth as a small fire flamed up. This gave him the chance he needed and he gradually piled nearby dead wood on until he had a large crackling fire. It was almost dark and there was one last major problem; was there enough wood to last the night?

If you've gathered wood in a forest you know there is usually plenty of old broken branches. After it snows it is very hard to find anything that will burn. Add to that the approaching darkness with no axe and you're almost back to a life crisis. He is able to find enough wood for the moment but he must go further from the campfire light each trip. It is now pitch dark and there is not enough wood to last half the night. My Palm Springs odyssey did not have such a single critical moment as this. My situation was more like a football game, steady and grueling. I would rather do the Palm Springs trip twice than face what Raymond was up against...

As he warms himself beside the fire he is now in extra innings. What to do? What to do? He thinks of places where there might be more wood. Without light he will be feeling his way around. The wood was hard enough to find in daylight. Remembering a dead tree that he had scouted earlier for branches, he realized there was plenty of wood, but no axe to cut it down. A plan is slowly forming as he locates the tree. It is not so large that he can't make it sway. Pushing it as hard as he can the tree begins to rock back and forth. A few more shoves and then he hears a crack; another home run. The tree is pulled over to the fire and the lower trunk thrown on. For the rest of the night as the fire burns down, more of the trunk is lifted on. As dawn breaks the last of the tree that saved Raymond's life is gone. Raymond heads out and quickly runs into a search party. The ordeal is over and hot coffee tastes especially good. It turned out that he was never more than a mile from his campsite during the entire ordeal.

Raymond and I came away with another life lesson. I hope this story is of value to other hikers. I still wonder about the people before me who were lost on the Palm Springs Mountain and were never found. Did they see the lights of the desert and feel they could not make it? Did they try to descend and get trapped? Were they injured in a fall?

Someday I would like to return to the top of Mt. San Jacinto, only this time with a camera instead of a rifle. It's a very beautiful part of our country. If it is evening and the sky is cloud-free you will see distant lights. If you are not in a hurry and if you watch carefully you might even be able to see them move.

I met Court Montrose at Avery Label. We were roomates and have been best friends ever since. Court cared for and managed Richard Palmer's estate in Whittier, CA. Palmer was the most important aircraft designer of the 1930's. He was the genius behind the Howard Hughes H-1 racing aircraft. It held the world speed record for many years and is hanging in the Smithsonian. He also influenced the design of many WWII aircraft.

My Children, Kim and Eric, were born a few years after I left Avery Label

CHAPTER 4

REFLECTIONS ON AVERY LABEL

In March of 1998 I read in the paper that Stan Avery, the founder of Avery Label Co., had died. The Avery name and labels can be found in every business office. In fact, I would venture to say there is an Avery product in just about every American home.

Stan started producing labels in his garage using an old washing machine to mix adhesives. I joined Avery Label Co. in Monrovia, California following my enlistment in the Army. My training in the Army as a Medical Lab Technician qualified me to work in Avery's Research Department. I worked on developing new formulations for pressure-sensitive adhesives, which in 1959 was as much an art form as a science.

I put on my job application the fact that I had won a Ping-Pong tournament. During my employment interview Norm Moline, my future boss, challenged me to a Ping-Pong match in Avery's recreation hall. Norm won and I went to work the next day. Working for Avery was my first exposure to the excitement of scientific research. I left five years later on a mission to help land a man on the moon.

Stan Avery believed in fundamental research. This kind of research may not produce profit for ten to fifteen years or even longer. The laboratory, unlike the rest of the company, operated with a very relaxed work policy. This was intended to promote creative thinking, as it was believed that research could not be hurried or scheduled. Inspirational ideas were more likely to occur if you felt at ease and not under pressure. Frequently the day would start with large informal gatherings over Avery's free coffee sometimes lasting till lunch and beyond. The subject of discussion was often current events, world problems, company gossip, etc. I remember the laughter as someone would break out singing a few bars of "Summertime." Occasionally a home repair project was brought in to work on. We never had to jump up and look busy when company management came by, but we did sometimes wonder what would constitute cause for dismissal. The only termination I recall was someone caught sleeping in his car several hours each day. Our research director rode a motorcycle to work (almost unheard of in those days). After work many of us could be found around the corner at "Fred's Club," where several hours of shoptalk would continue. Some of Avery's best new products were first designed on small napkins.

There developed a very close unity of purpose in the laboratory. Anyone struggling with a technical problem could expect unlimited help from associates.

Much of the serious side of Avery research involved constantly improving a variety of adhesive label products, resulting in colors and coatings for printing, and silicones for the backing paper. Mylar, foil, fabrics and exotic papers were but a few of the many choices Avery offered. We developed a double gum label that assisted colostomy patients and a high temperature label called "Heatex" that would withstand six hundred degrees. There was also a label for the eyebrows of a popular "Mattel" doll. Someone requested a label to stick on a real cow. (We didn't solve that one!)

Since Minnesota Mining was already producing Scotch Tape, Avery never considered competing for that market, as it would have only been marginally profitable.

There were times that we had trouble getting certain supplies. A chemical supplier informed us a resin in one of our adhesives was no longer available. When we inquired as to why, we found that the resin was derived from ancient tree stumps, which were now very rare.

There were two general categories of adhesives: removable and permanent. Permanent allowed manufacturers to label their products with serial numbers, how to use instructions, warnings, etc. The first permanent adhesive made by Avery was compounded from just three ingredients. For months various concentrations of the three ingredients were tested along with other possible additives. Surprisingly, after hundreds of tests, the very first experimental batch became the standard production formula. We often compared our work to that of "chefs" creating new recipes.

Avery also developed its own testing equipment. A standard 'tack' test used a highly polished stellite bar on which a measured label was pressed. The label was slowly lifted from the bar using a gradually increasing force. The force came from sand spilling into a beaker on a teeter-totter device. The more sand needed to lift the label, the more stick or tack in the adhesive.

My first major project was to develop a highly removable non-staining adhesive. This would result in a label that would adhere to a surface for several months and then could be easily peeled off without leaving a stain. One of many applications, besides retail pricing, for this adhesive was unexpectedly the "bumper sticker."

Early bumper stickers were virtually impossible to remove. The sun's UV rays and the high temperature of the metal usually welded most stickers to the bumper. Some bumpers would even need to be re-chromed to remove traces of the adhesive.

The sun also affected our removable adhesive, but instead of hardening, it turned gooey and could be rolled up in little balls. Not perfect, but superior to other products. The Democratic Party ordered several million bumper stickers. This was Avery's largest single label order at that time. Thirty years later, I attended a luncheon at the plant and received a standing ovation for this work. I was told over one hundred million dollars worth of the adhesive had been produced by Avery.

Avery Label Co. in the early 1960s began to expand. Plants were built overseas and new innovative products were to emerge. The Rotex label maker, which embosses letters on a plastic label strip, was developed at the same time Minnesota Mining produced its Dymo Tool. Minnesota Mining was Avery's closest label competitor and they were slightly ahead of Avery in the tool design. In the rush to get our Rotex product to the market place, I was given the urgent task of finding a suitable adhesive. The stiffness of the plastic strip meant the adhesive would need to be stronger than any adhesive previously developed. On a curved surface the rigid plastic strip tended to gradually straighten and lift off.

Luckily we located a suitable adhesive manufactured by Pittsburgh Plate Glass. Avery ordinarily developed its own adhesives, and never sought patents for its adhesive formulations. Patents were like advertising to your competitors, who could then develop a similar product.

Some Avery research did result in patents. I had received a patent for inventing a process called "Solvent Retention Adhesive." It was not a formulation for an adhesive, but a general adhesion method. A small amount of solvent would remain in the adhesive and produce the active mechanism for the label to adhere. The Solvent Retention Adhesive retained the convenience of a regular pressure-sensitive adhesive.

Pittsburgh Plate Glass was not about to divulge its own adhesive formulation. So Avery purchased the compounded adhesive from them and had it shipped from back East. This provided time for us to develop an in-house replacement. A senior chemist was subsequently hired because of his significant experience with adhesives. He was very knowledgeable in theoretical areas, but did not always appreciate Avery's more down to earth (but effective) research methods. Much of our success at Avery came from trial and error and intuition. The senior chemist had a temper and was especially hard on the junior members. This was not the Avery way and so it was with some sense of relief that he was promoted to Avery's new corporate research facility in Pasadena. There he proceeded to order numerous complex scientific instruments. As the deadline for Rotex production approached, we were informed that he had uncovered a weakness in our adhesive. He was using a spectrometer, which had detected an abundance of "free radicals" in our adhesive. A spectrometer measures very basic properties of matter on the atomic level. We had no idea what the relevance of this information was. He claimed that excess free radicals indicated the adhesive was inherently unstable and would slowly decompose.

I was called to a meeting in Pasadena that included all of the principals and Stan Avery. Tom Albright, a previous employee of Minnesota Mining, was sent with me to bolster our assertion that the adhesive was sound. We had used oven aging (one week in the oven = one year on the shelf) to prove the stability of the product. This was a common method used by industry; however, absolute verification of shelf life was impossible. Only time would tell.

I had very little to say at the meeting. Tom mentioned he could smell a particular resin in the adhesive. Our senior chemist emphatically assured us the spectrometer had detected no such resin. With little time to develop an alternative adhesive, Stan Avery correctly decided to take a chance. Deterioration of the adhesive was never a problem. When Pittsburgh Plate Glass stopped producing the adhesive they gave Avery their formula. The resin Tom smelled was in the formula. Thereafter, for years at Avery's board meetings, Tom's nose was mentioned!

The first production day for the Rotex label was well planned: Pink champagne, cake, etc. Anyone who was anyone at corporate Avery was present, including the Avery family.

As sometime happens with research, we had located a superior plastic film just a few weeks before the big Rotex demonstration. A West German firm sent us the new plastic by airfreight. Rushing to meet our deadline they may have hurried their production. The plastic rolls arrived a bit tacky. So the morning of the demonstration we carefully loosened the rolls and put them into a low heat oven to dry out. A technician, who on rare occasions used the same oven, came in that morning and without checking inside the oven, turned the heat up. Sometime later we began to notice an unusual smell. Someone yelled "OVEN!" We found the plastic rolls looking like ice cream in July. The technician couldn't be fired. He was, however, in much more danger of being taken out in the back of the plant.

So much for the demonstration, but "oh" the champagne and cake were still great!

Apollo being rolled out

Apollo 14 - 1-31-1971

Dr. John Houbolt, showing his unique approach to Apollo

Courtesy: NASA Archives

CHAPTER 5

NORTH AMERICAN APOLLO

After graduation in 1965, with a degree in Math from Los Angeles State College (now California State University at Los Angeles), I made a decision that would forever redirect my life. Spotting an opportunity on the job board at school, I set my sights on outer space. My work at Avery Label was secure, but was it in any way comparable to the excitement of Apollo? Don't ask--the decision was beyond my control. I accepted a job at North American Space and Information in Downey, California.

My parents and I were having an ongoing discussion about the feasibility of going to the moon. My father, having been born when the first airplane flew, could not imagine how such a feat could be possible. I remember reading the classic Life magazine article that suggested travel to the moon might be possible in twenty-five years. The article described a spacecraft that was to be assembled in space. It looked like a series of tanks strapped together with no consideration for aerodynamics. This was some of the first information the American public was to read about the near perfect vacuum of outer space. It was a wonderful medium for high-speed flight, but considered uncertain as an environment for human beings. Sputnik and our first satellites along with President Kennedy's proclamation to go to the moon by the end of the decade seemed to alter my parents' view of space travel.

The only calculating device that I remember using at the time was the slide rule. Most of the math problems in school contained relatively small whole numbers because hand held calculators were not yet in use. More complex math functions were found in a book of mathematical tables. I remember it was very distracting to stop in the middle of a problem and interpolate in a table. My trigonometry book had a third of the pages devoted to columns of numbers. In spite of the fact that these numbers are now available at the touch of a button, lower division mathematics is little changed in substance today.

My computer career started on the most powerful computer of its day, the IBM 7094. North American Space and Information in Downey had two of them and they were busy day and night. The company was like a small city, and I would walk the halls and factory bays each chance I could. I never got to see it all. The first program I wrote contained one page of FORTRAN. This turned out to be the closest scrape I ever had with a programming bug. The engineer I was helping found a mistake in one of the equations I coded. This was a simulation for Sub and Super circular equilibrium glide lines, which are entry angle tolerances for the returning Apollo capsule to be restrained with. During the critical descent stage the earth's atmosphere gradually slows the fiery return until deploying parachutes can be used safely. There are two dangerous outcomes if a narrow corridor along the flight path is exceeded.

When the astronauts return to earth they must be careful not to dig too severely into the atmosphere, thus generating high G forces (they could be injured). On the other hand, if they come in with a shallow angle they could skip away from the earth, and when their orbit returned them their oxygen would be gone.

When I certified the program I mentioned to the division boss, Rube, the close call I had with the bug. He shrugged it off and told me he had spilled a cup of coffee into a 1401 (an early IBM business computer).

"Zero defects" was NASA's motto for Apollo, and we all were constantly reminded about the dangers of space flight. I felt a tremendous responsibility towards the astronauts in spite of my exceedingly obscure role. I would never again complete a program with a bug; I am now sixty years old, so perhaps there is still time to break my perfect record.

We were located in a converted building that had been a WWII combat aircraft factory. My office for a while was pressed against the roof of the old manufacturing bay and the ceiling was full of odd sized pipes. If you weren't careful you would bump your head on one of them, and we all did it at least once. The door was built like a bank safe, in case top secret work was underway.

This was the largest single NASA facility (32,000 employees). Manpower had been quickly amassed for Apollo and was increasing at a prodigious rate. Within six months most newly hired engineers were managing large groups. Funding seemed to have no limit, and the motto was "If two could not complete a task then hire ten."

I never knew why, but they moved our offices every month or two. In the evening before you left for home you would tape up your desk and file cabinets. The next morning you would find them in a different building. Everyone went through this over and over. There was a joke going around that the reason for this was because there wasn't enough room for everyone and they had to have one group on the move at all times.

In the production facilities there was an incredible collection of gigantic machines for manufacturing and testing the Apollo capsules. There were about twenty-five spacecraft in various stages of assembly in one building. Some capsules were assembled just to train the technicians. There were large "bench presses" you could stand up inside, towering mock-ups of the Saturn boosters and the largest clean room in the free world. This was a very exciting period in history.

Over lunchtime we did one of two things: play chess or read computer manuals. I must say my chess game improved considerably with my brown bag competitors. I also started playing with a rated player in the evenings. The intensity of chess can be considerable. A hundred years ago Army generals were expected to play chess because it was believed their battle field strategy would improve. As my computer programming ability improved I found a corresponding improvement with my board play.

I soon realized I was unintentionally memorizing every move of the match. I would analyze my strategy and ponder the good and bad moves for hours. I had trouble concentrating on anything else. No matter what I was trying to think about my mind would gravitate to the game. This began to interfere with work and sleep. Unable to conquer this compulsion, I had to stop playing chess even though I enjoyed the game very much.

For me, orbit determination is like chess. It has been thirty years since those programming days, but a day rarely passes without thinking about my work at JPL. Oh, those incredible second order differential equations!

During the intervening years I have learned more about my compulsive behavior. Having read the book The Boy Who Couldn't Stop Washing, written by Judith L. Rapoport, I came to understand my compulsions. I clearly had the characteristics of moderate obsessive-compulsive behavior, but was not even close to the extreme examples of dysfunctional disorders. In fact I began to see that it was one of my strengths. I am sure that checking and rechecking a computer program was the only way to wring the last bug out.

A good friend of mine is an obsessive-compulsive person. He is perhaps more extreme than I am. He gets sores on his hands from washing them so often. He opens doors with his elbows or uses a tissue when he must touch a handle. He wears gloves all day. I am not nearly this bad, but I do step over cracks and have learned to fight the urge to check and recheck the door I just locked.

The one year I worked at North American was spent converting older programs to a new FORTRAN II. Many of these involved celestial mechanics. The main work was on the Massachusetts Institute of Technology (MIT) guidance scheme. We attempted to verify that a planned orbit transfer during a motor burn could be accomplished. MIT developed the analysis using something called the "V to Go" vector. This was the vector difference between the probe's instantaneous velocity and the velocity of the intended orbit. Acceleration (burn) was directed (aimed) to drive the "V to Go" vector to zero. This proved to be a powerful mathematical method for executing orbit transfers.

I found it interesting that my technical boss would study the stock market two or three hours each day. This was an opportunity for me to learn from him. I began to grasp the technical charts and believed I could master them. I took a bath in the market. It would be thirty-five years before I bought another share of stock and then I took another bath.

The Apollo capsules were tested in a variety of ways. For instance, a return capsule was suspended by a chain and splashed into a tank of water. The aluminum honeycomb heat protection shield came loose. It was back to the drawing boards for one of the many technical groups. It was not unusual to see open expanses of a hundred separate desks for the Journeyman Engineers. Thin open-air partitions were randomly spaced around the bays for the managers. My immediate supervisor would look out over the top of this partition several times each day. He reminded us of the cartoon figure the WWII GI's scribbled on walls with the caption "Kilroy was here." We were never sure what he was looking for, but with his fingers on the partition and only his forehead and eyes showing we nicknamed him "Kilroy." He was an extremely shy man who was uncomfortable to speak with. He was very competent technically but had zero leadership ability. The rapid expansion of the workforce often placed individuals in these situations. Several years later the book The Peter Principle, written by Laurence J. Peter and Raymond Hull, would explain the widespread occurrence of this circumstance.

The Saturn rocket was the largest and most powerful ever constructed. The payload that escaped the pull of Earth's gravity was a complex segmented group of modules. The general details for this design were the product of a single engineering genius, Dr John Houbolt. In a memo that he described as "a voice crying in the wilderness," he detailed a Lunar module, lander and return capsule configuration that would radically depart from conventional thinking. Initially shelved and in competition with Wernher Von Braun's Earth orbiting plan, his insight helped overturn the Buck Roger's direct ascent "everything goes, everything returns" design. It significantly reduced the payload weight and turned out to be the only practical design given the power of the launch vehicles. The separate functionality of these independent piggybacked modules may have saved the Apollo 13 crew.

Far and away the most disturbing moment was when the astronauts were killed testing the re-entry vehicle at Cape Kennedy. Later I saw a capsule that had scorch marks near the door. This had a brief depressing effect on us all, but we continued to look ahead. The thought of walking on the moon was not going to be delayed. Many changes, especially in the oxygen system, would be necessary. I believe Kennedy's dream had real meaning for all of us and we re-doubled our efforts.

North American was a large and impersonal company but to a person we were all in a sense unified. I do not believe the Russian competition was paramount. This was America at its finest.

President Kennedy was gone but through us his dream would live.

STRING OF PEARLS

Mariner 4 was launched on November 28, 1964 and journeyed 228 days to the Red Planet, providing the first close-range images of Mars.

Courtesy Of: NASA/JPL/CALTEC

CHAPTER 6

STRING OF PEARLS

It was not really a very distinctive photo. Most people on the street would think it was probably just another Surveyor picture of the moon. The rocky rubble between the irregular sized craters suggested a surface exposed to the haphazard bombardment of meteors, and little else. How could this be the most important picture of the Twentieth Century? There was no artistic grace. If anything it looked drab, forbidding and even boring until you realized it was the first close-up picture of Mars.

The Mariner IV took the photo from several thousand miles above the Martian surface. A video camera recorded the image and a radio transmitter sent the digitized picture to a receiving station on Earth. I do not believe that any single picture has ever transcended so many human cognitive visions. The Jet Propulsion Laboratory in Pasadena California released it to the media and shortly thereafter the most fantastic place beyond the confines of Earth was revealed. Jules Verne wrote books, Percival Lowell saw "canals" and Buck Rogers altered public opinion, but in 1965 Mariner forever bound our reality. In a single moment centuries of wondering vanished as we adjusted to this technological feat.

Video images are comprised of tiny shaded spots or "pixels." These are obtained from the light entering the lens of a video camera. In a serial manner they are "analogized" by the scanning electron gun and are represented by an electric current. Ultimately stored on magnetic tape, the image of Mars was translated into a series of numbers where each number determined a "shade of gray" for a pixel. Every number is then digitized as a group of bits or on/off switches. This is all in turn consistent with binary computer circuitries.

That fateful day in 1965 a radio wave carrying this image of Mars left the Mariner transmitter and headed for Earth at the speed of light. A number of minutes would elapse before the first bits arrived at the telemetry station antenna on Earth. Surprisingly at that exact moment the last bit was leaving the proximity of Mars. This was not an engineering requirement for this remarkable photograph to be stretched out between Earth and Mars for a split second. Unnoticed, this fact would remain concealed as the world contemplated the unprecedented event.

The first man and woman must have looked into the night sky and wondered what those lustrous starry objects were. I'm sure early pilgrims climbed the mountains hoping for a better view. Galileo with his telescope most likely strained his eyes to get a better incarnation of this wonder.

On Earth that day, that momentous moment so many of us waited for was about to happen. A few more minutes to wait and then the computer enhancement revealed a strip of the first frame. While the remainder of the picture was processed eager eyes looked for the first time at the surface of Mars. There was conclusive evidence to settle the age-old question... craters and definitely not canals. To the anxious scientists at JPL the lack of canals was not surprising. I believe that many of us held onto a last private hope that when the astronomer,

Giovanni Schiaparelli, in 1877 first described these structures, he was actually describing water ways.

A string of pearls (those beautiful bits) were connecting us to our sister planet for a brief moment but as lightning passes across the sky they were immediately hurled Earthward. These taciturn pearls would shortly attract world attention but to the engineers at JPL their salient consequences would be almost inexpressible.

The following four pages are among many thousands of pages of technical writing associated with the DPODP. "Solar Pressure" "Motor Burn" "Numerical Methods" and "Relativity" are shown for illustrative purposes and are courtesy of NASA/JPL/CALTEC:

or, for the components,

$$\ddot{r}_{br} = [a_r + b_r(t - T_{AC1}) + c_r(t - T_{AC1})^2][u(t - T_{AC1}) - u(t - T_{AC2})]$$
$$+ \Delta a_r + \frac{c_1 A_p}{m r_{SP}^2}[G_r + G'_r(\measuredangle EPS) + \Delta G_r]u^*(t - T_{SRP})$$

$$\ddot{r}_{bx} = [a_x + b_x(t - T_{AC1}) + c_x(t - T_{AC1})^2][u(t - T_{AC1}) - u(t - T_{AC2})]$$
$$+ \Delta a_x + \frac{c_1 A_p}{m r_{SP}^2}[G_x + G'_x(\measuredangle EPS) + \Delta G_x]u^*(t - T_{SRP})$$

$$\ddot{r}_{by} = [a_y + b_y(t - T_{AC1}) + c_y(t - T_{AC1})^2][u(t - T_{AC1}) - u(t - T_{AC2})]$$
$$+ \Delta a_y + \frac{c_1 A_p}{m r_{SP}^2}[G_y + G'_y(\measuredangle EPS) + \Delta G_y]u^*(t - T_{SRP})$$

The terms in these equations are defined as follows:

$\hat{U}_{SP}$ = a unit vector directed from sun to spacecraft (spacecraft roll axis)

$\hat{X}^*$ = a unit vector in spacecraft $+X$-axis direction (spacecraft pitch axis)

$\hat{Y}^*$ = a unit vector in spacecraft $+Y$-axis direction (spacecraft yaw axis)

$\hat{U}_{SP}, \hat{X}^*, \hat{Y}^*$ forms a right-handed, orthogonal, spacecraft-fixed coordinate system; thus, $\hat{U}_{SP} = \hat{X}^* \times \hat{Y}^*$

a_i, b_i, c_i (where $i = r, x, y$) = solve-for coefficients of low-thrust acceleration polynomials, km/s^2, km/s^3, km/s^4

t = ephemeris time

T_{AC1}, T_{AC2} = epochs at which attitude-control acceleration polynomials are turned on and off, respectively; epochs may be specified in UTC, ST, or A.1 time scales (not E.T.)

$u(t - T_{AC1}) = \begin{cases} 1 \text{ for } t \geq T_{AC1} \\ 0 \text{ for } t < T_{AC1} \end{cases} \quad T_{AC1} \to T_{AC2}$

$\Delta a = (\Delta a_r, \Delta a_x, \Delta a_y)$ = input (a priori) acceleration, km/s^2 (value of each Δa_i, $i = r, x, y$, will be obtained by linear interpolation between input points on any time scale)

$c_1 = \frac{J A_E^2}{c} \frac{1 \text{ km}^2}{10^6 \text{m}^2} = 1.031 \times 10^8 \frac{\text{km}^2 \text{kg}}{\text{s}^2 \text{m}^2}$

where

J = solar radiation constant
 = 1.383×10^3 W/km^2
 = 1.383×10^3 kg/s^3

$A_E = 1.496 \times 10^8$ km (mean distance earth–sun = 1 AU)

$c = 2.997925 \times 10^5$ km/s (speed of light)

A_p = nominal area of spacecraft projected onto plane normal to sun–spacecraft line, m^2

m = instantaneous mass of spacecraft

r_{SP} = distance from sun to spacecraft

T_{SRP} = epoch at which acceleration from solar radiation pressure is turned on (becomes effective); epoch may be specified in UTC, ST, or A.1 time scales (see Glossary for time scales UTC and ST)

$u^*(t - T_{SRP}) = \begin{cases} 1 \text{ for } t \geq T_{SRP} \text{ and if spacecraft is in sunlight} \\ 0 \text{ for } t < T_{SRP} \text{ or if spacecraft is in shadow} \end{cases}$

G_r = solve-for effective area of acceleration of spacecraft in radial direction from solar radiation pressure divided by nominal area A_p

TECHNICAL MEMORANDUM 33-451

"instantaneous" motor burn (see below), is given in Ref. 7, (p. 30) as

$$\ddot{\mathbf{r}}(MB) = a\hat{\mathbf{U}} \{\mu(t - T_0) - \mu(t - T_f)\} \quad (220)$$

where

a = magnitude of $\ddot{\mathbf{r}}(MB)$ vs time

$\hat{\mathbf{U}}$ = unit vector in direction of $\ddot{\mathbf{r}}(MB)$ vs time

T_0 = effective start time of motor, E.T. value of the solve-for UTC epoch, T_0 (UTC)

T_f = effective stop time of motor, E.T.

t = ephemeris time

$$\mu(t - T_0) = \begin{cases} 1 \text{ for } t \geq T_0 \\ 0 \text{ for } t < T_0 \end{cases} \quad T_0 \to T_f$$

The effective stop time T_f is given by

$$T_f = T_0 + T \quad (221)$$

where T is the only solve-for burn time of the motor in ephemeris time.

The acceleration magnitude a (in km/s²) is given by

$$a = \frac{F(t)}{m(t)} C$$

$$= \frac{F_0 + F_1 t + F_2 t^2 + F_3 t^3 + F_4 t^4}{m_0 - \dot{M}_0 t - \frac{1}{2}\dot{M}_1 t^2 - \frac{1}{3}\dot{M}_2 t^3 - \frac{1}{4}\dot{M}_3 t^4} C \quad (222)$$

where[22]

$F(t)$ = magnitude of thrust at time t (polynomial coefficients F_0, F_1, F_2, F_3, and F_4 are solve-for parameters)

---
[22]It should be noted that the coefficients

$$\dot{M}_0, \frac{1}{2}\dot{M}_1, \frac{1}{3}\dot{M}_2, \frac{1}{4}\dot{M}_3$$

are actually the coefficients of a Taylor series; that is,

$$\dot{M}_0 = \dot{M}_0, \quad \frac{1}{2}\dot{M}_1 = \frac{1}{2!}\ddot{M}_0, \quad \frac{1}{3}\dot{M}_2 = \frac{1}{3!}\dddot{M}_0, \quad \frac{1}{4}\dot{M}_3 = \frac{1}{4!}\ddddot{M}_0$$

The coefficients of the polynomial (Eq. 222) appear as in the first of these equations because they are supplied in this form by the Propulsion Division at JPL.

$m(t)$ = spacecraft mass at time t

$C = 0.001$ for F in newtons and mass in kilograms

m_0 = mass of spacecraft at T_0 in ephemeris time

$\dot{M}_0, \dot{M}_1, \dot{M}_2, \dot{M}_3$ = polynomial coefficients for propellant mass flow rate (positive) at time t, $M(t) = \dot{M}_0 + \dot{M}_1 t + \dot{M}_2 t^2 + \dot{M}_3 t^3$ (coefficients are not solve-for parameters, but must be supplied)

$t = \text{E.T.} - T_0 (\text{E.T.})$, s

where E.T. is seconds of ephemeris time from January 1, 0 h E.T., 1950.

The unit vector $\hat{\mathbf{U}}$ of thrust is given by

$$\hat{\mathbf{U}} = \begin{pmatrix} U_x \\ U_y \\ U_z \end{pmatrix} = \begin{pmatrix} \cos\delta \cos\alpha \\ \cos\delta \sin\alpha \\ \sin\delta \end{pmatrix} \quad (223)$$

where

α = right ascension of $\hat{\mathbf{U}}$

δ = declination of $\hat{\mathbf{U}}$

given by

$$\left. \begin{array}{l} \alpha = \alpha_0 + \alpha_1 t + \alpha_2 t^2 + \alpha_3 t^3 + \alpha_4 t^4 \\ \delta = \delta_0 + \delta_1 t + \delta_2 t^2 + \delta_3 t^3 + \delta_4 t^4 \end{array} \right\} \quad (224)$$

where the polynomial coefficients are solve-for parameters. As an example of how these coefficients may be solved for, a midcourse maneuver will be considered. In this case, an impulsive velocity increment $\Delta\dot{\mathbf{r}}$ is assumed; a midcourse maneuver program computes the roll turns and burn time so that the computed $\Delta\dot{\mathbf{r}}$ is obtained by the spacecraft. Given

$$\Delta\dot{\mathbf{r}} = \begin{pmatrix} \Delta\dot{x} \\ \Delta\dot{y} \\ \Delta\dot{z} \end{pmatrix} \quad (225)$$

∂z/∂q and the weights applied to each residual, along with the *a priori* parameter vector and its covariance matrix (Ref. 17, p. 32), one computes the differential correction Δq to the parameter vector.

Starting with q and Δq, one computes a new spacecraft ephemeris, residuals, and partial derivatives, and a second differential correction is obtained. This process is repeated until convergence occurs and the weighted sum of squares of residual errors between observed and computed quantities is minimized (Ref. 18, p. 24).

Differential correction is applied for two conceptually different purposes. One application considers the effect of errors in the observational data themselves. For the other application, let it be supposed that a preliminary trajectory has been computed on the basis of some simplified physical model (e.g., assume only two-body forces) or on some simplified trajectory (e.g., assume a hyperbolic or elliptic trajectory). Even if perfect observational data are given, subsequent observations would not necessarily agree with what they were computed to be on the basis of the preliminary trajectory.

The quantities that are usually differentially corrected are *a priori* estimates made to minimize the sum of weighted squares of residual errors between observed and computed quantities. These estimates include injection parameters, physical constants (implying that they are not actually constants; that is, their numerical values are subject to improvement—e.g., the astronomical unit or the speed of light), maneuver parameters, and station locations.

A. Interpolation and Differential Correction of Basic Planetary Ephemerides

Predictions of the motion of celestial bodies can be presented in either of two forms: (1) as general but complicated formulas, with time as argument, from which position at any epoch can be computed; or (2) as tables listing discrete, prespecified epochs, from which positions at other than tabular epochs can be obtained. These tables are called ephemerides.

It has become customary to rely exclusively on ephemerides for astronomical work involving lunar and planetary motion because the labor required for the preparation of an ephemeris can be allocated to the solution of many different problems.

1. Interpolation. The basic planetary ephemeris data consist of predictions of lunar and planetary positions and of the corresponding velocities. The ephemeris data are usually given in heliocentric coordinates for the planets and the earth–moon barycenter and in geocentric coordinates for the moon. However, coordinates referred to any of the bodies as center may be obtained by a translation of centers (see Section VI). As the planetary position ephemerides are tabulated at 4-day intervals (an exception is Mercury, whose data are given in 2-day steps) and the lunar ephemeris at ½-day intervals on a standard ephemeris tape used at JPL, it is necessary to use an interpolation scheme to obtain intermediate values of positions and velocities. An Everett's formula that uses second and fourth differences is usually employed for the positions and velocities; the formula used (Ref. 19, p. 273) is given by

$$x(T_j) = ux_0 + tx_1 + \frac{u(u^2-1)}{3!}\Delta^2_{m0}x_0 + \frac{t(t^2-1)}{3!}\Delta^2_{m1}x_1$$

$$+ \frac{u(u^2-1)(u^2-4)}{5!}\Delta^4_{m0}x_0$$

$$+ \frac{t(t^2-1)(t^2-4)}{5!}\Delta^4_{m1}x_1 \qquad (384)$$

where

T_j = desired Julian date, $T_i \leq T_j < T_i + h$

h = step size of data

T_i = point in time at which data are tabulated

$t = \frac{(T_j - T_i)}{h}, \quad 0 \leq t \leq 1$

$u = 1 - t$

$x_0 = x(T_i)$

$x_1 = x(T_i + h)$

Δ^n_{mi} = nth modified difference

The modified differences are intended to facilitate the use of Everett's fifth-order interpolation formula by "throwing back" sixth- and eighth-order differences on the second and fourth differences:

$$\Delta^2_{mi} = \delta^2_i + a_{26}\delta^6_i + a_{28}\delta^8_i \qquad (385)$$

$$\Delta^4_{mi} = \delta^4_i + a_{46}\delta^6_i + a_{48}\delta^8_i \qquad (386)$$

which are to be treated relativistically are specified by input. This manner of treating relativity spheres eliminates any discontinuities in the integration of the equations of motion due to relativity.

In the early formulation of the general theory of relativity, the equations of motion for a massless particle moving in the gravitational field of other bodies were taken to be the equations of a geodesic. That is, the motion of a particle was obtained by solving the field equations for the metric tensor, which describes the geometrical properties of space and time, and by assuming that the particle follows a geodesic curve in this geometry. The actual method for determining the motion of a system of n heavy bodies directly from the field equations was obtained for the first time by Einstein, Infeld, and Hoffmann in 1938. This method, which is referred to as the EIH approximation method, is, according to Bazański (Ref. 14, pp. 13–29), in principle, the only tool for obtaining an approximate solution to the problem of the motion of n heavy bodies in the general theory of relativity.

Table 1. Radii of relativity spheres[a]

Celestial body	Mean distance from sun a_p, AU	Sun–planet mass ratio μ_S/μ_p	Radius of relativity sphere r_p, km × 10^6
Mercury	0.387	6,000,000	2
Venus	0.723	408,500	7
Earth	1.000	333,000	9
Mars	1.524	3,100,000	4
Jupiter	5.20	1,047	400
Saturn	9.55	3,502	300
Uranus	19.20	22,900	200
Neptune	30.1	19,300	200
Pluto	39.5	360,000	50
Moon	1.000	27,100,000	1

[a]Moyer, T. D., JPL internal document, Jan. 4, 1968.

From Infeld's equations of motion, after some computations and simplifications,[26] the resultant equation for the relativistic acceleration of body i "due to body j," denoted by $\ddot{\mathbf{r}}_i(j)$, is

$$\ddot{\mathbf{r}}_i(j) = \frac{\mu_j(\mathbf{r}_j - \mathbf{r}_i)}{r_{ij}^3} \left\{ -\frac{4}{c^2}\phi_i - \frac{1}{c^2}\phi_j + \left(\frac{\dot{s}_i}{c}\right)^2 + 2\left(\frac{\dot{s}_j}{c}\right)^2 - \frac{4}{c^2}\dot{\mathbf{r}}_i \cdot \dot{\mathbf{r}}_j - \frac{3}{2c^2}\left[\frac{(\mathbf{r}_i - \mathbf{r}_j) \cdot \dot{\mathbf{r}}_j}{r_{ij}}\right]^2 + \frac{1}{2c^2}(\mathbf{r}_j - \mathbf{r}_i) \cdot \ddot{\mathbf{r}}_j \right\}$$

$$+ \frac{1}{c^2}\frac{\mu_j}{r_{ij}^3}[(\mathbf{r}_i - \mathbf{r}_j) \cdot (4\dot{\mathbf{r}}_i - 3\dot{\mathbf{r}}_j)](\dot{\mathbf{r}}_i - \dot{\mathbf{r}}_j) + \frac{7}{2c^2}\frac{\mu_j \ddot{\mathbf{r}}_j}{r_{ij}} \qquad (254)$$

where

r_{ij} = coordinate distance between bodies i and j

$(\dot{s}_i)^2, (\dot{s}_j)^2$ = square of velocity of bodies i and j, respectively

ϕ_i = Newtonian potential at body i

ϕ_j = Newtonian potential at body j

and

$$\ddot{\mathbf{r}}_j = \sum_{m \neq j} \frac{\mu_m(\mathbf{r}_m - \mathbf{r}_j)}{r_{mj}^3}$$

The acceleration of body i "due to body j" is a function of the position and velocity of bodies i and j and the positions of all other bodies, which contribute to the Newtonian potential at bodies i and j and affect the acceleration of body j (terms 7 and 9 of Eq. 254). Although the effects of other bodies are included, all terms are proportional to μ_j and hence are attributable to the presence of body j. The effect of the mass of body i on its own acceleration is contained in term 2 (its contribution to the Newtonian potential at body j) and in its contribution to the acceleration of body j (terms 7 and 9).

[26]Moyer, T. D., JPL internal document, Jan. 4, 1968, and Khatib, A. R., JPL internal document, Feb. 11, 1969.

CHAPTER 7

JPL HISTORY OF DPODP-DPTRAJ, 1966-1971

The Double Precision Orbit Determination Program, DPODP (later referred to as the ODP), is a computer program containing nearly 300,000 FORTRAN statements. It is divided by function into two separate portions. The first is Double Precision Trajectory program, DPTRAJ, which numerically integrates the equations of motion of a space probe. The second, also referred to as DPODP (synecdoche), differentially corrects and solves for the a priori[1] parameters, which influence a spacecraft's trajectory.

The Telemetry data collected by the Deep Space Net (DSN) at JPL contains about a dozen observable types, principally Doppler and Range. "Covariance matrices with a weighted least square regression analysis" will correct the parameters influencing the spacecraft's motion. The residual error between the observed and computed (light-time solution)[2] observables is minimized as the constraining factor used to "determine the orbit."

The DPODP-DPTRAJ was certified for mission use in 1970. This was the culmination of seven years work at JPL with the immediate goal of supporting Viking and Voyager projects. The first operational test was with Mariner VI and VII. NASA had been anxious to complete this program because ongoing parallel efforts on other navigation projects throughout the country were providing unnecessary duplication. DPODP was to become the navigation backbone of our exploration of outer space. A consequence of this work was also more accurate determination of thousands of parameters concerning the physics of our solar system. I was told Einstein's general theory of relativity was confirmed. In addition, this program reconciled the work of Kepler, Newton, Gauss and other physicists of historical note.

I believe I was the only computer analyst who worked on both the DPODP and DPTRAJ portions. I contributed one third of the FORTRAN statements and was especially productive in helping adapt the models[3] supplied by the engineering group into forms which could be

---

[1] A Priori - the best deductive values available, either self-evident, reasoned, or a presupposition.

[2] Light-time Solution, or LEG as it is also called, is the mathematical process leading to the calculation of the observables from the numerically integrated trajectory of DPTRAJ. The residual difference is a subtraction that assumes or supposes consistent units. The position and velocity vectors determined by DPTRAJ are a mathematical construct stretching in some cases for millions of miles and existing at a single instant in time. The observables are recorded at a tracking station. At the moment they are collected they are not spread instantaneously over time and distance. Because they are recorded at an antenna on the earth's surface they don't represent a vector quantity and subsequently have been transmitted at the finite speed of light. The light time solution is a recursive process of translating the numerically integrated trajectory vectors to the equivalent observables so the residual subtraction can be effected. Because the origins of these identical quantities are dissimilar their difference is important and essential as the controlling mechanism of orbit determination.

[3] Models are generically defined as any collection of mathematical statements (usually equations) with a purpose. The burn model of DPTRAJ is an example. Polynomial coefficients describe how the rocket warms up

programmed. There were about five hundred equations in our models. The numerical solution of the equations of motion for a spacecraft was my most satisfying accomplishment.

Following Sputnik there had been a frantic period of orbit tracking. Telescopes and cannon mounts were used to locate the first satellites. The "tack and string"[4] elliptical methods were enlarged in scale to room size proportions. JPL eventually produced the Single Precision Orbit Determination Program SPODP. Two stages of development evolved separate programs with different parameter sets (maximum fifty).

These programs (SPODP) were advanced for their era and suitable for solitary planet flybys such as the Mariner projects, but orders of magnitude additional accuracy would be required in the near future. NASA's biological quarantine stipulated accuracies of less than one chance in ten thousand of an accidental collision during a Mars landing or continuing parking orbits for at least twenty years. Mid-course maneuver fuel limitations also eliminated the single precision program from any realistic contribution to the "Grand Tour" Voyager mission. An enormous leap in accuracy was imperative. In the early 1960s the Thompson-Ramo-Wooldridge Corporation (TRW), a major California aerospace and defense contractor, with forty PhD's, had attempted a large-scale effort and failed. JPL decided to take on this daunting task in-house.

As previously noted, I began working at the North American Space and Information Division in 1965 verifying the MIT (Massachusetts Institute of Technology) Guidance Scheme for Apollo. Mike Warner hired me to work at JPL in 1966 as a contract employee for a software company called Informatics. Mike was the only analyst who was available from the old SPODP project and was given responsibility for managing the DPODP software development at its inception.

By the time I joined the DPODP project, FORTRAN coding had been cautiously underway for a few months. There were also several major technical hurtles that were unresolved, computer capacity and numerical integration to name a couple.

A year earlier there had been a dispute about the efficacy of coding without adequate documentation. Fred Lesh, Mike's boss, had previously delayed all coding for a year so preliminary documentation could be completed. The original staff of thirty programmers working on the project during my first year at JPL was mainly International Business Machines (IBM) Federal Systems contractors. The difficulties in coordinating this large group were

---

after ignition with start and stop times being a critical determination. The polynomial functional values describe the force on the probe from this thrust. The word "model" may be better understood by considering an airplane model. This proportionate construction using metal, plastic or wood is not the same as a real airplane. Similarly an applied math model provides an approximate numerical value of the subject described but is rarely if ever an exact representation.

[4] An ellipse is the set of all points such that the sum of their distance from two fixed points is a constant. If you make a length of string this constant and fix both ends of the string with tacks, you can trace out an ellipse by stretching the string out with your finger. Scaled down, this is a primitive approximation of the Earth-Sun system. The Sun is one tack and the Earth is your finger slipping along the loop of string. Fred Lesh mentioned to me that this relic of desperation was still stored in a dusty room somewhere at JPL.

soon evident by the slow pace of progress. A staff of five zealots would eventually conclude the work.

The DPODP was destined to become the largest application program in the world. The second generation IBM 7094, the largest mainframe available at the time, was understood to be inadequate because of processor speed and addressable storage. It was originally necessary to overlay hundreds of subroutines in a very unrealistic structure while waiting for larger third generation equipment. This old war horse 7094 did, however, allow us a couple of years of restricted program development while waiting for the new computers.

DPTRAJ was the first program converted to the third generation UNIVAC 1108 Exec 8 and was actually used by UNIVAC as a diagnostic tool when they experienced serious difficulties at JPL with their first operational system. An associate, Al Joseph, realized the new computer system would probably not be reliable, especially when the anticipated heavy usage at JPL started. With this incredible insight we desperately worked around the clock when the ll08 was first barely operational. This led to successfully converting DPTRAJ months ahead of any other programs. Nearly a year would pass before the 1108 was functioning reliably enough for any effective program development throughout JPL. The IBM 360 had similar difficulties meeting contractual schedules.

The DPODP was eventually converted to the IBM 360. The UNIVAC 1108 was the primary computer for all software development connected to the DPODP. Because of serious mission deadlines and lengthy delays in completing the DPODP we were given unrestricted use of JPL's computer facilities. It did seem extravagant at the time, a fully dedicated IBM 7094 and later a UNIVAC 1108 for up to eight hours of daily prime time use. This continued for many months, and I believe the amount of money that I individually spent for computer time on this application program will never be equaled.

With the Grand Tour opportunity and the subsequent Voyager mission came realizations that navigation accuracy could not be attained without radical state of the art improvements. The somewhat relaxed requirements for targeting closest approach for a single planet encounter would be intensified by orders of magnitude with a multi-planet slingshot energy transfer. The numbers of parameters in the gravitational models would increase a hundred fold.

Each model describing the unbalanced forces on the probe would receive a dramatic mathematical boost.

1. Solar Pressure (Sunlight carries energy and momentum $E=MC^2$), Eleven - year solar cycle, Umbra-Penumbra shadow models

2. Plasma Drag (Cosmic particles and protons)

3. Attitude Control leaks (non-coupled jets) Locked on the bright star Canopus.

4 Direct Gravitation for all Planets with options to differentially correct and solve for osculating orbital elements (Brower and Clemens, Set #5) within the
JPL Planetary ephemeris tape.

PATHWAYS TO THE PLANETS

    5   Planetary and Moon Oblateness

    6   Indirect Oblateness for Earth's moon (accelerated coordinate frame)

    7   Instantaneous Maneuvers (Solar Panel Deployment, spring separations)

    8   Motor burn

    9   Planetary Atmospheres

    10 MASCONS[5] (massive chunks of mostly iron 50 km across buried beneath the surface of the moon)

Ted Moyer modeled general relativity as an unbalanced force. Schwartzshield (metric tensor) Euler-Lagrange (equation), Infeld, de Sitter and Einstein were scientists he referenced for his analysis. This was the one model I never really got a seat of the pants feeling for, not that I didn't try. I do remember reading books on relativity.[6] A couple of times I remember feeling a warm, wonderful sensation indicating I finally understood relativity, but it usually only lasted 5 or 10 seconds. So I trusted that Ted had done the correct thing.

The more complex relativistic coordinate systems might have been employed, but the Cartesian X-Y-Z Newtonian frame made the balance of the models much more comprehensible. Since relativity was treated as a correction, we were probably referred to as "Correctionists"[7] in relativity circles. Without allowing for relativity, additional mid-course maneuvers would jeopardize the fuel available for the Voyager.

There was a diverse list of upgrades. Doubling the precision of the models would place a strain on the supporting library functions ordinarily provided by the computer operating system. Routines like Sine and Cosine had to be reformulated. A new approach to numerical integration would be necessary. Mathematical formulations, theoretically generating rooms full of telemetry tape, required optimization.

With thousands of parameters to be possibly "solved for,"[8] it was decided that in a single case no more than fifty could be corrected at a time. Twenty additional parameters were

---

[5] MASCONS may have been partially responsible for overshooting the landing site of the first Apollo landing.

[6] Einstein presented his theory of Special Relativity in 1905 and then extended this to General Relativity published in 1916. General Relativity included principles that explained discrepancies in Newton's Law of Gravity. The force used in Newton's inverse square law would become the curvature of a relativistic space around a massive body. A new geometry had emerged and additional analysis would be required to adopt these ideas into the DPODP. Einstein had produced a momentous first step, but the requirements of planetary navigation would combine motion, time and electro-magnetic signals into a single comprehensive problem. This Einstein could not have envisioned. Completing this analysis fell squarely on the shoulders of Ted Moyer.

[7] Correctionists is a tag placed on those who choose to visualize the universe as fundamentally Newtonian and treat relativity as a small aberrance. The distortion of space, the slowing of time and the bending of light do not conform to our notions of a stable Euclidean world. Our intellect of necessity prefers Newtonian physics and I for one was gratified to know that relativity could be placed in little convenient boxes throughout the analysis. Correcting for relativity meant I could safely live in my square xyz world. I still envy those rare individuals who are comfortable in Einstein's world. I wonder if they still see the sun go down like I do?

designated "consider"[9] because their improved values would be withheld.[10] Fortunately the householder[11] method allowed us to avoid inverting a 70x70 double precision covariance matrix to accomplish the regression. Dave Hilt managed this last crucial modification of the DPODP segment. This 'solve for' vs. 'consider' flexibility allowed studies of the interactions of correlated parameters. It was interesting that the gravitational acceleration of the sun and solar radiation pressure were exactly opposite forces and could not be solved together in the same case.

Telemetry data processing was going to require additional engineering. We needed to add additional Doppler types; antenna angle offsets for station locations, ionospheric refraction, tracking by ship, landed spacecraft observables, relativistic corrections to the light time solution and relativistic corrections to time keeping frequencies. Time and timekeeping is a fundamental principle of physics and the dominant centric influence on virtually all processes within the DPODP. Time is a consequence of motion and the equations that represent motion are driven by time, i.e. the so-called independent variable of the equations of motion. Telescopes observing stars, the swing of a pendulum, or the vibration of an atomic nucleus are used for time reckoning. In each case motion is connected to an observer and affected by a relativistic reference frame, the gravitational field. The non-uniformity of time must be taken into account for precision navigation. Time and its influence on space travel is an active area of investigation and standards for measuring it change as technology advances. (E.T., A.1., UT0, UT1, UT2, UTC, and S.T. were all systems of time used by the DPODP).[12]

In the entire DPODP the largest sector of analysis involves the equations of motion. No matter how thoroughly modeled, unaccounted secular forces[13] on the probe will eventually cause a

---

[8] "Solved for" parameters appear as constants within the analysis. Often they are coefficients of polynomials and they all influence the equations of motion or the observable calculations. The mass of the sun (Gm or gravitational constant for the sun) is an example of a solved for parameter. The orbit determination process improves the value of the sun's mass so that the numerically integrated trajectory better fits the telemetry data.

[9] "Consider" parameters are the same as solved for parameters except their values are not differentially corrected (changed) during a particular case. This allows studies of the selective effects certain parameters have on the orbit determination process. "Exactly constrained" parameters are functionally related (tied together by fundamental physics) and the correction process adheres to this constraint. "Inexactly constrained" parameters are similar to exactly constrained parameters except their functional relation may be weighted.

[10] By withholding corrections to certain parameters on an experimental basis a view of the complexities of regression analysis is possible. Minimizing the residuals and selectively allocating the corrections gives insight into normally distributed statistical error and the weighted least square process.

[11] The Householder method was introduced into the DPODP as a replacement for a matrix inversion. After the equations of motion have been solved.... after the variational equations have been solved....after the partial derivatives for the observables have been computed....the last step planned for each observable was to invert a 70X70 matrix. There are a variety of mathematical methods that theoretically would accomplish this last step and the coding is not complicated. However, there would emerge several critical problems connected to this seemingly unavoidable inversion. If you took the computer time for one inversion and multiplied it by the millions of observables likely to be processed during a mission, you came up with an unacceptable number. In addition, the numerical nature of the off-diagonal elements indicated many of them approached but were not equal to zero. This created the possibility of accuracy dilution during the inversion process. Luckily, before a lot of effort had been spent investigating the options, householder came to the rescue.

nominal (preferred) trajectory to wander. Periodic motor burns used to adjust the trajectory are at the heart of navigating between the planets. A committee determines the frequency of mid-course maneuvers rather than any established numerical criteria. Although mid-course maneuvers would compensate for navigational deficiencies, conserving fuel was a dominant factor in abandoning the single precision program, SPODP. I never saw a line of its code. Its internal structure had very limited influence on me. However, it represented a significant historical moment on the pathway to higher precision navigation and the exploration of the solar system.

During low points in the DPODP development, when schedules had to be slipped a year or more, outspoken critics perceived that it was too large, too complex, too expensive or even impossible. One disgruntled engineer resigned because he lost his patience with the software progress. He circulated an open memo that expressed his anger. He was sure the project would never be completed and equally sure his career was damaged because he had to wait so many months. Mike's experience with the SPODP left him with an unwavering determination that we could finish it.

As the physics of celestial mechanics was advancing, the supporting mathematics needed to stay abreast of the physics. Numerical techniques used to solve the equations of motion were improved. With round off and truncation error the controlling factors, starting procedures, step size efficiencies, forward and reverse integration techniques were among some of the improvements.

Experience with the Apollo project left me with several impressions about numerical integration packages in general. They were usually associated with the most complex portions of a program. There were some well-established traditions concerning predictor corrector methods, starting procedure instability, halving and doubling of step sizes, etc. I wondered if these entrenched principles should be reassessed. The Cowell[14] choice for the central body gravitation strategy I found comforting. The first and second sum difference

---

[12] Ephemeris Time (E.T.) appears as an independent variable in the equations that describe motion. It is theoretically the uniform but un-measurable argument for the ephemerities (motion records) of celestial objects. Ephemeris time is easily understood as an abstract concept and verbose within the mathematics of celestial mechanics. It is however impossible to observe or measure directly and represents conceptually the ideal (smooth) time. Atomic Time (A.1.) is based on a frequency from oscillations of a nuclear standard. Cesium was changed to hydrogen as the standard for atomic time in the 1970s. A.1. represents the best estimate of ephemeris time but is influenced by fluctuations inherent in the Earth-Sun relativistic field. Universal Time (UT) is the precise measure of time used for civil timekeeping and several categories are registered and are obtained by optics from meridian transits of stars connected to the Earth's rotation. Irregular rotation and polar wanderings of the earth distinguish UT0 and UT1. UT2 is approximately A.1. and when broadcast is referred to as UTC. UTC is called Greenwich civil time and is broadcast by the National Bureau of Standards. Station Time (ST) comes from oscillations of a rubidium atomic clock on site at a tracking station. Two "solve for" parameters connected to time keeping were the time difference (in seconds) between A.1 and E.T. at 1958.0 and a correction of the cesium frequency.

[13] Secular forces are unbalanced forces on the probe not included as part of any model and assumed to emerge no matter how exhaustive the analysis. Inferences suggest they originate from some as yet to be discovered source. It is highly unlikely any system of models can ever completely describe celestial motion. It is understood that even if all the forces on the probe were modeled exactly as they exist within the solar system, other galactic sources would eventually cause a probe to stray from its prescribed path.

equation[15] was the choice for the numerical structure (Everett's). Here was an opportunity to review some pre-conceived notions about integration. Dr Lawson strongly suggested that additional analysis was needed in this area. Since we did not require flexibility to solve general groups of second order non-linear differential equations, we could focus on our own specific problem.[16] Second order non-linear differential equations are the workhorse of our modern technological world. Why second order, I haven't a clue. The overall navigation problem is a two-point boundary value problem. The second point is left to peripheral search routines (drivers) not associated with the DPODP. The first point, referred to as injection conditions (the state vector[17] of the probe at the beginning of a mission), allowed the option of thirty-six different coordinate systems. Joe Witt and John Ekelund were the cognizant programmers.

I discovered that the two-body Euler-Taylor[18] series approximations for initiating the starting procedure were unnecessary and subsequently could be removed. Simply projecting the initial condition accelerations (from the derivative box) backward in time as a first approximation proved to be feasible. It surprised me when I found it required no extra iterations to converge (suggesting stability). In fact, I had difficulty finding an example, no matter how irregular, where the starting procedure failed to converge. This had not been the case in instances of other projects I had been involved with. Following the rule and in keeping with the high requirements for accuracy within the numerical integration, no step forward in time was allowed without a fully converged backward difference line.

---

[14] The Cowell method structured the second order differential equation describing point source gravity in the most sensible way possible. Using Cartesian coordinates each force on the probe including the strongest gravitational source (central body) was depicted as a separate contribution to the overall acceleration of the probe. Only because there are alternate approaches to the problem is the Cowell method distinguished by a name. Enke uses a conic solution for the central body gravity and the rest of the forces are similar to Cowell. Calculus of Variations integrates a non-Cartesian coordinate system such as the elliptic elements.

[15] A difference equation is formulated with stepwise differences instead of continuous variables. A sequence of consecutive higher order differences is referred to as a difference line. When hand tabulated these differences often form one side of a triangle, i.e. a line. Higher order differences are generated in the DPODP by a succession of subtractions connected to discrete fixed intervals. The probe accelerations over ten time steps are differenced (subtracted) then these differences are subtracted, etc. Each complete set of differences produces a single contribution to the difference line. The first and second sums represent the position and velocity in a fashion conforming to the set of interpolating coefficients. A difference line facilitates a convenient polynomial interpolation process, i.e. position and velocity at an arbitrary time.

[16] The decision not to require a generic numerical integration package was based on manifold considerations. The early design had been hasty and engineering specifications were not then clearly established. The ongoing process of program modifications eventually buried any hope for an independent package. The numerical integration became a tool for state of the art investigations and any generic restrictions would have slowed the progress. In the end the mathematical knot had been tied so tight that only a complete overhaul lasting perhaps years would separate DPTRAJ from the numerical processes. Note: This is exactly what happened.

[17] The state vector, of the probe is seven numbers which determine completely the probes motion at a moment in time. The chosen inertial frame that is used is the earth mean equator and equinox of 1950.0. In Cartesian coordinates this is X, Y, Z, Xdot, Ydot, Zdot. Although there are many equivalent coordinate systems (36 in the DPODP), seven numbers is a minimum requirement.

The traditions of halving the step size during integration I found to be somewhat arbitrary. Venerated traditions within the world of mathematics were re-examined in light of revolutionary advances in computers. Techniques, often promoting the name of a famous mathematician, expounded upon many decades earlier were found to be inconsistent with the new more powerful technology. Mathematics and electronics, energized by aerospace, were both in rapid states of development. With millions of calculations easily and conveniently available to analysts, new approaches to numerical problems could be pursued. If truncation error indicated a step size reduction, why is one half chosen? Is it because halving is a tradition emphasized in many numerical integration packages?

The last retained difference was utilized to select a better and usually larger step size reduction and was consistent with accuracy requirements. This saved unnecessary starting procedure iterations (and computer time). The extreme change in acceleration caused by a motor burn would seem to present a problem due to the sheer magnitude of the thrust. It was the largest force on the probe by far. The automatic step size control of the numerical integration package could have been challenged due to the near discontinuous change in acceleration. During long missions in solar cruise the step size might be extended to several days. The step size for a typical motor burn would be a couple of seconds. It was decided to put in a suggested starting step size for each motor burn (This was the only model to artificially alter the step size). This override and the inherent stability of the polynomials used to describe the warm up of the burn chamber worked very well. Instantaneous maneuvers were discontinuous in position and velocity and were not integrated across. A normal re-start after the maneuver was required.

Additional innovations were developed which advanced traditional Adams-Moulton Predictor/Corrector techniques. (This was later referred to as the Adams Type). A 'Predictor only' method worked well but truncation error forced a smaller step size. This did approach equivalent accuracy after the corrector cycle. Reduced derivative calculations were only marginally achieved. 'Predict-pseudo correct' was like 'Predictor only' except in the case of large derivatives such as a gravitational central body or motor burn these derivatives were re-calculated on the corrector cycle. Pseudo was a bit more efficient than 'Predictor only' and gave us insight into the process of numerical integration. An option to select the number of differences to integrate over, possibly up to ten, was found not especially useful. All I/O[19] files assumed ten differences anyway and so data storage could not be reduced.

The difference lines from the equations of motion always exhibited a puzzling characteristic. The ordinary pattern for the higher differences was a diminishment in magnitude. Somewhere around the seventh or eighth difference the pattern would reverse. This growth could be

---

[18] The two-body Euler-Taylor series was a clumsy conglomeration of mathematical estimates, which projected the state vector of the probe to other time points. Newton's two-body recursive method would have been a better choice for this task from an accuracy standpoint, except it would have required far more coding (IBM 7094 core storage had to be conserved at all cost). The Euler extrapolation formula was very simplistic and a few terms of the Taylor series joined this crude approximation. It was inserted into the program as a desperate attempt to save storage. The only good thing that could be said was that it seemed to do the job.

[19] I/O (input/output) files are usually tape, drum or disc storage for data input to the program and/or data output by the program. I/O is also referred to as mass storage and is always connected with some mechanical device.

explained as resulting from round-off error. Although the interpolating coefficients would probably render these spurious differences harmless, JPL analysts decided we should disallow them no matter how insignificant they may be. I never felt that I fully understood the source of this pattern. I attempted to independently verify how this error worked its way into the difference line. This was a very tedious and unsatisfying exercise due to the large numbers of calculations involved. I was resigned to be a self-nominated compulsive worrier and sometimes pre-occupied with minutia.

According to Dr Lawson, president of the National Numerical Analysis Society, the numerical integration package in DPTRAJ was the most comprehensive available anywhere in NASA. Traditional Adams-Moulton was first referred to as Adams-Type because of our work.

Rich Brodie helped write the very first numerical integration package for DPTRAJ. Because of the massive changes involved in my work, Rich would probably not recognize any part of it. Yet, because of his early work and since I never had time to redesign, his fingerprints are all over it.

Tracking station locations needed to be determined to within a few inches. This was accurate enough to theoretically sense continental drift. Spacecraft locations near the planets were accurate to tens of kilometers. By the mid-1970s it was estimated that 10^{15} bits of data had been processed by the DPODP without a single bit of error. Lack of a single program bug represented unheard of reliability for a large program.

Einstein's Theory of Relativity was verified. This is perhaps the only verification using somewhat ordinary conditions (on planet earth). The speed of a spacecraft, (tens of thousands of miles per hour) is a small fraction of the speed of light. A comparison theory termed Brans-Dicke was evaluated but unsubstantiated. Einstein's work with general relativity held. (Of course, we were all pulling for Einstein).

One tenant of orbit determination is that "everything works" or "nothing works." The final results are an extraction of information from an enormous reservoir of data. A small mistake in the analysis or the programming could reduce the accuracy of all results. There was never a guide or set of parameters to be used as a goal to establish benchmarks indicating the entire program was checked-out! This was leading-edge navigation. So-called 'hand checks'[20] could only be performed on very small isolated sections of the program. A complete hand check on one step of the numerical integration was impractical. To converge one difference line (ten steps prior to injection) requires more than a million calculations. A complete hand-check on one step was impossible. Using very clean structured coding made de-bugging somewhat easier (although "hack and slash" was a term we used for our numerical methods).

There were no formal certification guidelines at this time. An innovative group of test cases were developed (Al Khatib, the cognizant engineer). I worked closely with John

---

[20] Hand checks were "pad and pen" calculations supported by a hand held calculator. These were useful because they did not involve the computer or the coding they were verifying and therefore represented an independent validation. The section of coding selected for a hand check was usually augmented with temporary FORTRAN write statements which generated printout used to compare with the hand calculations.

Ekelund because the variational equations (from Sturms and Moyer) were synchronized to my numerical step size controls. Partial derivatives of the state vector with respect to the spacecraft parameters were Ekelund's responsibility. He assumed my error control from the numerical integration was sufficient for his integration procedure. To save re-calculation of the unbalanced forces I made the derivative box from the equations of motion available to his link, VARY, through an I/O file.

There was an alternate method for analytically solving the variational differential equations. The variational equations were as elegant and formally precise as the equations of motion. Deep within the formalism of mathematics other descriptions of the same principles were to emerge. There are two main approaches to calculus--Leibnitz and Newton--and both essentially lead to the same results. So it was that a second equally appropriate method leading to the variational solutions was possible.

This alternate method of differencing using perturbed and unperturbed analysis did proceed theoretically to the same partial derivative solutions. Since the mathematical analysis was significantly different, but the results were approximately the same, the perturbed method could be used to verify the analytic method. The issues were actually over the "infinitesimals," and what occurs when they become "finite." To make the applied math even more ponderous there were several approaches, all leading to differenced perturbed partial derivatives.

The method, chosen against the wishes of Earl Grossnickle, was very complicated from a programming standpoint. Earl was a top analyst from the single precision program and was assigned to the perturbed Link called PDQ. (VARY was the analytical Link) The name PDQ seemed to have been selected with a complete disregard for the computer work required. Or, perhaps it was called PDQ because the general mathematical statement of the generic differential equation is relatively simple. He left the lab before completing the work.

A couple of years later Mike gave me this task, and as I picked up his uncompleted program I soon realized a simpler approach would be possible from a programming standpoint. It occurred to me that it would be feasible to utilize the DPTRAJ numerical integration package to re-run the nominal case with a perturbed parameter. This was going to harness the strength of the computer to give us what we wanted and represent a classic example of the incalculably small becoming finite. Each perturbed parameter would generate a different trajectory, i.e. different partial derivatives. The differencing, without the complexity of the numerical overburden in VARY would be relatively simple to program.

I contacted Earl Grossnickle at Lockheed and he agreed with my supposition. He said politics had dictated his original approach and he was aware an easier method might be possible. He felt my ideas were valid and Mike accepted this new hypothesis. PDQ did not require efficiency since it would never be used operationally; its only purpose was to verify the analytical method. At least three man-years of effort were saved.

I believe Earl had left JPL over this dispute. In the evenings at the Informatics offices in Los Angeles, Earl and I had worked on a proposal for the DPODP contract. This would have given Informatics a sizable management role in the entire project. Earl had all the credentials and experience to manage the entire project, and freedom from political pressure was obviously to his liking. The proposal we worked on was very large and detailed. No other company, bidding on this contract, could have come close to presenting the issues the way we did (we were both insiders).

JPL rejected all of the proposals. This once and for all decided the case; we (JPL and me) were going to finish it. I can only surmise that they had finally decided to do it in-house no matter what. Possibly Earl was still out of graces with someone... I have no idea.

Now, how to bring everything together and finish the job? Without a clear set of guidelines to use as a goal it would remain unclear what criteria would determine completion of the work. We were pushing accuracies beyond any "here to fore." We couldn't throw a party and declare the program finished either. This last struggle would require our finest efforts. I began to develop what I would call back door methods. Restricting DPTRAJ, so as to compare with other programs, I started with a two body, Newton-Raphson, closed form solution (Al Joseph's conic experience was helpful). Then I compared DPTRAJ with the programs that generated the JPL planetary ephemeris tape. These programs were leading edge for planetary motion and were equivalent in accuracy. I put the probe where a planet should be and derived comparable results. The SPODP was helpful for simple cases.

Forward and reverse integration options were interspersed throughout DPTRAJ. If you can integrate backward in time and regain your injection conditions you should obtain a great deal of confidence in your forward results. NASA-Ames and Houston were now in contact with me. Post processing of later Apollo flights was an early operational use of the DPODP. Top level JPL managers were flying to Washington regularly to reassure NASA that the work was being completed. Mike Warner, Al Khatib and Al Joseph insulated me from the political arena of funding. I was free to pursue anything and everything with unanimity. Although there had been numerous schedule slips throughout the entire DPODP project, I never missed a software delivery deadline.

I was constantly re-deriving the same differential equations, tracing the same coding path through the triggering sequence, confirming and then reconfirming over and over. I uncovered bugs in subroutines left to us by programmers long gone but hardly responsible considering the reckless programming controls of those early days. They had been part of the early flush of coding which now only added to landfills. Some bugs would require a month or more to locate. I usually carried a small deck of IBM cards in my pocket to think about. At the bank, at the cafeteria, I never knew when or where, sometimes in the middle of the night "bingo," I would have an answer. I kept a pad and pen beside the bed so I could write them down without waking completely up. Without explanation I would run my hand across a two hundred-page listing and open it to the exact line of code causing a problem!

There was a constant stream of revisions from the engineering group, a nightmare for most programmers. They just blended in with modifications ongoing in the numerical integration package. I was contributing more to the engineering. The new analysis would arrive in memo form, usually from the cognizant engineer Al Khatib. The logical process from the general analysis to the minute details of the FORTRAN coding was now entirely my responsibility. I had learned to expect some engineering shortcomings in the complete enumeration of a problem.

A typical example of this was the allowable range of angle parameters. It was not unusual to be given equations that were valid only if the angles were restricted to the first quadrant. Assuming these equations translated to other quadrants was dangerous. I could brace the engineers about the other quadrants or the coordinate axis and often uncover inconsistencies. As the program began to jell I became very protective of the solidarity demonstrated by our programming group. We had arrived and overcome adversity, challenging new mathematical mountains. It was now apparent that an emerging world of mathematics existed between the high level textbook equations of celestial mechanics and the bit diddling of a modern digital computer.

I was continually finding bugs (over two hundred major corrections the last year of work) and wondered how I would know when the last one was discovered. Many of these corrections altered substantially the state vector of the probe and would doom any mission. I remember subroutine CEPHEM, which interpolated the planetary ephemeris file: I had to change the coding one statement at a time over several years. It seemed to be "drop by drop . . .day after day . . .when would it end?"

For several years my mind worked actively twenty-four hours a day on the DPODP. Driving down the L.A. freeways I would trace out the connections between the differential equations and the coding. Over and over, searched for some small crack, some discord in the logic. I was a classic "obsessive compulsive" person. I washed my hands, stepped over cracks and then in my mind checked and rechecked the reasoning a hundred times each day. I sometimes would emerge from this cloud and find myself in the gardens at JPL hardly remembering how I got there. I finally began to realize it was all going to end and that the final certification would probably be a cakewalk.

Al Khatib orchestrated the formal process that continued around the clock for several days. He had arranged for a fully devoted UNIVAC 1108 during these sessions and asked me to be available. Most of the problems were in the operating system environment. Al and I had a couple of early morning breakfasts on him, but I had a strong feeling it was finished. Mike Warner's strong leadership had always been there; he never wavered in his conviction that we would eventually win. No more bugs were showing up. The math, the coding, everything was ready. The hardware could now be launched into space and the years of waiting were ahead of us.

With the combined efforts of over one hundred analysts and programmers this was their victory, and the MOST important ingredient had been an unyielding dedication to TRUTH.

THE LAST MARINER PARTY

New JPL Director (1954): Dr. WIlliam Pickering

Courtesy Of: NASA/JPL/CALTEC

CHAPTER 8

THE LAST MARINER PARTY

(My wife typed this, and the extra comments belong to her.)

Naturally it being the 1960s, there was bound to be some kind of incident involving marijuana. Slade was our inter-lab mail carrier, and his girl friend worked for IBM. I recall on her desk was a plaque that stated "I Blow Marijuana." It was superimposed over an IBM logo and must have irked her superiors. There was a rather remarkable incident regarding Slade. He was the first male at the Lab that I recall having shoulder length hair. I was possibly the second, but it was not as long as Slade's hair. Slade was a goofy string bean sort of kid. I remember one day he was totally out of breath; half-laughing and half excited, he said he had been chased down the hall by Dr. Pickering who was the JPL director at the time. I found this absolutely bizarre. He said Pickering had been trying to find out who had the long hair. Slade, of course, had to deliver mail occasionally down the hall where the director's office was located. Someone alerted him to watch out because his long hair might get him into trouble. As the story goes, Slade spotted Pickering at one end of the hall. Quickly Slade went the other way. Pickering speeded up after him and Slade took off running. Now Pickering was running. Of course, being faster afoot he managed to get away.

The Mariner parties began as a celebration of a successful mission. The tradition did not take long to become popular. Since it was not located at JPL, but at someone's home, everyone could relax. Employees were usually looking for ways to unwind because there is always a very tense period just before planetary encounters. This party had all the trappings of a memorable event: cars were double-parked, crowds in the street and all the things that go with large parties.

Back in the 1960s marijuana was still fairly unusual in many social settings. Somehow marijuana was brought to the party, but it was not a pot party. (THEN WHAT WOULD YOU CALL IT?) I can only give you a second hand account, as I was not there. (OH SURE) I would not be surprised, though, if it were Slade who had something to do with everything going awry. Very respectable people (OH REALLY?) were enjoying the excitement.

I don't know how it happened that the police were alerted, but they raided the party. This was a time when serious jail sentences were possible for just one joint. One of our sharpest DPODP engineers was among those arrested. There was general mayhem at the time of the arrest, with people trying to get away. The Mariner project manager was there; he got scared, went out the backdoor, climbed the back fence, and got away. He had nothing to do with the marijuana, but he could see that there was going to be a messy controversy. The story was on the front page of the Pasadena Star News, a scandal getting first class press coverage. Dr. Pickering had more to worry about than long hair, and it was "THE" last Mariner party.

Although in the beginning I had supported President Johnson's decision to become involved in Vietnam, I was beginning to have second thoughts. Driving to work one morning I passed a large group of high school students in Pasadena carrying big signs. That was later recognized as the first Vietnam protest in the country. John Ekelund had briefly mentioned that we might not be doing the right thing in Vietnam when we started bombing. He also warned that the government might not be divulging the whole story. This began years of doubt and mistrust for me.

Soon I was one of the early outspoken Vietnam protesters. There was a great deal of contention at the lab about the war. In addition to the war controversies there were other complex political issues. I remember Al Joseph coming in one morning; he was absolutely infuriated. Police had shot up a house in the Los Angeles area of Watts. They had killed people inside while searching for Black Panthers. It turned out to be a private school. Things were just getting out of hand in the political world. What with my staunch viewpoints and possibly because of my new long hair "hippie" look, someone placed a sign on my office door that read "Super Dove." I left the sign there for several months.

UNIVAC 1108

Courtesy of Charles Babbage Institue, University of Minnesota MN

IBM 7094

Courtesy: IBM Archives

**Mechanical Device for trajectories
before computers took over.**

Courtesy Of: NASA/JPL/CALTEC

CHAPTER 9

UNIVAC VS. IBM

In the beginning man first counted on his fingers. Our base ten system attests to that. Then later came mechanical computing devices like the Abacus. The actual historical evolution of the modern digital computer is far too complex for me to delineate. It does, however, represent the finest example of the unlimited capacity of the human intellect.

My mother's cousin invented the pop-up toaster. He approached my Uncle, Jelly Smith, the business mind in our family, with an offer to go into business selling the toasters. Jelly had started several businesses in the small Minnesota town of Lancaster. After some thought Jelly decided he wasn't interested. He didn't figure they would sell. The Toastmaster Company was started without Uncle Jelly.

About a dozen relatives of mine have received a degree in mathematics. Several have medical degrees. This can all be traced to my great-grandmother, who emigrated from Sweden to northwest Minnesota traveling from the East Coast in a covered wagon. Her occupation was midwife.

Throughout high school the only calculating I remember was with pencil and paper. I used a slide rule most of the time in college. Occasionally I used a Marchant mechanical calculator at Citrus Junior College in Azusa. The first computer I remember seeing was during a visit to Harvey Mudd College. A Chemistry teacher was using one to study carbon sub oxide. The first computer I programmed was in 1965. It was the IBM 7094 Direct Couple at the North American Space and Information Division in Downey, California.

After many years of study, these are the impressions I have of the people who blazed the trail to the modern world of information technology. There are three truly momentous breakthroughs that stand out. First came Charles Babbage, because his gears were real digital units on a large scale. The next was Alan Turing, because his engine showed the limitless possibilities for binary computing. Finally John Von Neumann, because of his description of a stored program, in non-mechanical electronic circuits, which was exactly the model used for all future digital computers.

The following is the first couple of pages of a ten-page guide to FORTRAN that I wrote in 1980 for my Physics students at Humboldt State University:

"Guide to FORTRAN Programming"

There are two classifications for electronic computers: analog and digital. Historically older, the analog computer makes use of directly measurable quantities such as current and voltage in order to perform arithmetic calculations. Because there are fundamental limitations with the accurate measurement of these quantities, the analog computer is only used in limited applications. It is the digital computer that is responsible for most of the advances in computing occurring in the past two decades. Since the

1940s there have been well defined breakthroughs in the computing capacity of the digital computer. Each succeeding generation produced a computer with dramatically increased computing power.

1950s First Generation - Vacuum tubes

1960s Second Generation - Integrated circuits

    (transistors)

1970s Third Generation - Silicone chips

    (transistors)

Although technology advances are continually occurring, there is no well-defined fourth generation. Instead, dramatic reductions in computer hardware costs have been accomplished with mini-computers and microprocessors.

The fundamental unit of information in the digital computer has two states - on and off (analogous to a light switch.)

Consider how a group of three on/off switches or bits may be used to represent binary numbers. (1 may be arbitrarily considered "on" and 0 considered "off ")

	Bits			Decimal number
Binary number	0	0	0	0
Binary number	1	0	0	1
Binary number	0	1	0	2
Binary number	1	1	0	3
Binary number	0	0	1	4
Binary number	1	0	1	5
Binary number	0	1	1	6
Binary number	1	1	1	7

Bits are grouped into computer words. Each computer word has the same number of bits and may be used to represent a single number. A computer word may also be used as an instruction for the computer to execute an arithmetic calculation.

John Von Newman first stated the concept of an electronic digital computer. He observed that arithmetic calculations (addition, subtraction, multiplication, and division) could be performed with electronic circuitry.

Programs are numbers as well as arithmetic instructions, which are stored electronically in the memory of the computer. The computer keeps track of which computer words are "numbers" and which are "instructions." The instructions are executed (or allowed to continue) one by one. Groups of instructions may be executed repeatedly but no more than one instruction is executed at one time. No moving parts are necessary for the arithmetic instructions.

The calculations will take place with the speed of the electronic pulses. As important results are obtained, they may be output from the computer electronics to a printer or other device with moving parts. For some programs, many millions of arithmetic calculations will occur without moving parts.

Example of computing as a logical process:

    Input - Submit program to computer

(deck of cards read by card reader)

    Processing - Calculations executed

       (central processing unit in control)

Output - Results of calculations (printer

    producing hard copy)

During early development of computers, programming was a tedious process. Programs were complex to develop, since the programmer was often working on the bit (or electronic) level. In order to simplify the programming process higher level languages were developed. With these languages it was unnecessary to understand the electronic complexity of the calculations in order to program the computer.

The IBM 7040-7094 Direct Couple was the joining of two main frames. One computer handled I/O and the other was principally for number crunching. The operating system was batch mode. Double precision used two single precision floating-point words each with separate characteristics and mantissas. I will never forget this single precision augmentation because of the work necessary to unpack both words from the octal dump in order to find the decimal number. True double precision with a single characteristic and mantissa would have to generally wait for third generation computers. Some second-generation systems, like Honeywell, did eventually get hybridized to match this third generation capability. The IBM 360, a truly third generation system, had a single characteristic and mantissa for its double precision word as did the UNIVAC 1108. Relating to the DPODP accuracy requirements, I had investigated algebraically increasing the precision beyond double precision for a few key calculations. I did not find the additional accuracy particularly useful and double precision remained adequate for our purposes.

For the IBM 7094 user, cards and tape were the principal input. Punched cards, tape and printed listings were the output. JPL operated five IBM 7090 or 7040-7094 direct couple computers at the Pasadena facility. The directly addressable core contained small ferrite rings with three wires passing through each of them. There were thirty two thousand single precision words of which about eight thousand were used by the operating system. This core size was the single most critical deficiency of the computer. Early DPODP work used a FORTRAN II compiler with two very minor assembly language routines. FORTRAN in those days was not machine independent and the assembly language restricted the portability of the program.

The operating system required only two cards: the IBJOB card, where user identification, printer and execution time limits were punched; and the IBFTC card, used to separate each FORTRAN subroutine for the compiler. Keypunch machines were available near our offices with optional courier service to the computers. If we were allocated a dedicated block of computer time it was usually in the SFOF/DSN facility. Daily routine usage was at a 7094 located near my office. Late at night if the computer operator was in a good mood I could wait for my runs inside the computer clean room (always a pleasant temperature in the summer). High-speed printers always fascinated me because of their knotty entanglement and elaborate metal-on-metal clatter.

Because of heavy computer traffic during the average workday, I would rotate my work with a second deck of IBM cards so I could be constantly busy. I had asked Mike Warner to give me at least two tasks at all times. I was usually waiting on one run submission while working on the other. Mornings were always exciting, since you could expect two "presents" waiting for you in your box. On a good day you could hope for three runs per task. Weekends you expected a gusher, getting in ten or twelve runs a day. On the other hand, if your deck came back in a bag that meant someone had dropped it. You could then only hope that you had updated your sequence numbers recently. Tape integrity was always a problem. We would make three backup copies for our important tapes and sometimes all three would turn out to be inexplicably unreadable.

Keypunch machines created a very unusual problem. The confetti-like pieces of the card that had been punched out had to be disposed of carefully. These tiny sharp rectangular pieces were extremely dangerous if they got into your eye. At IBM if you were caught playing with the card confetti you would be summarily fired.

Most Houston rocket launches were in the morning. The only launch that I attempted to watch was Mariner VIII on closed circuit TV at JPL and it was destroyed over the ocean. The cause of the problem was eventually determined to be the software. The coding mistake was traced to two incorrectly punched holes in a computer card. After that disappointment I stayed away from watching mission operations preferring the quietness of my office, a windowless room. The work was classified and someone told me windows were a breach of regulations. I didn't exactly understand this, since JPL had very tight security throughout the grounds. The lack of windows in my office was just as well, since the sight of smog always troubled me.

The only coffee available was from those button laded machines (very poor by today's standards). I drank six to seven cups a day. There were always good and bad days. One day, after having lunch at the cafeteria, my eyes swelled shut. Turned out it was the clam chowder.

I usually used "green sheets" to submit my first coding of a subroutine to a keypunch operator (I don't type). Then I one-fingered all the corrections (Forty years later, I am writing this book the same way). I found I could analyze coding for hours sitting at a keypunch machine. JPL maintained its own in house system group that coordinated modifications with corporate IBM. This is the only 7094 installation where IBM allowed its operating system to be modified by customer personnel.

The JPL system group was very competent and was extremely helpful with our system problems. It was a regular occurrence to find that we had exceeded the capacity of an obscure and unusual feature of the system. I was just as likely to be battling the computer system as the DPODP program. I became aware of the methodology a compiler used. I do not believe writing in assembly language would have saved any storage since I was fully anticipating the vagaries of the compiler. The coding was being compressed to save core and I was just as likely to be working on a core dump as the FORTRAN listing.

Reformulate equations, optimize coding and complex overlays were constantly being scrutinized and juggled to free up storage. Each time I was able to incorporate a clever piece of coding to create more storage, I felt a warm, wonderful sense of well being. The following day, however, I would probably be pinned against the storage wall again.

I could have removed all assembly language from DPODP with a FORTRAN compiler trick I knew that would jump you into executing a data set. Mike and I decided there was no valid reason to remove this trivial amount of assembly language, which was used to unpack the bits used for triggers.

IBM sold ninety percent of the world's computers during the 1960s. Any programmer with any sense wanted IBM because it was simply the best. The whole operation of Big Blue was indefatigable, with IBM dominating every aspect of the world's computing. In the 1960s this appeared destined to continue for decades. In all likelihood it would have if it weren't for a decision, most likely at the congressional level, to break the monopoly and spread the wealth of this increasingly important technology. Not promulgated through anti-trust laws, the decision turned on government agencies giving contracts to other computer manufacturers even if their equipment was inferior. The Navy gave UNIVAC purchase orders for their soon to be operational third generation 1108 exec 8.

JPL also acquiesced. The new IBM 360 would be used only for real time mission communications. The UNIVAC would be our baby. The only advantage UNIVAC apparently held over IBM was that their operating system was designed from the ground up. IBM always seemed to have business characteristics embedded in all their systems even when the users were largely scientists. This was of course, understandable considering the history of IBM. The basic blueprints for the UNIVAC 1108 Exec 8 system were sound but the big, big problem was that the early hardware and software did not approach the reliability we expected from IBM equipment.

The transition period gave some of us an uneasy feeling of impending doom. IBM was no longer going to support the 7094. We were handed its operating system and wished good luck. This began an interim period where the JPL systems group would patch the 7094 system to meet our immediate requirements. It was of some help, especially in freeing up a small amount of core. This band-aid merging of DPODP and the 7094 system lasted about six months. With some regret and no fanfare the last operational 7094 was packed up and sent to a university. We should have retained the IBM 7094 until the UNIVAC 1108 had demonstrated it was capable of supporting our work.

Alva Joseph now enters the picture. I have never known anyone who understands and anticipates computer problems better than Al. From the scantest of information he was able to correctly predict that an extended period of time would pass before the 1108 would be fully operational. This turned out to be a full year beyond the scheduled delivery date. Had JPL management heeded this staggering insight, an exceedingly anguished episode in the history of JPL could have been avoided. I don't believe JPL ever again lost a full year of software development.

The roots of this dilemma are not completely clear to me. The second and third generation computers required large climate controlled rooms. Without an interim period of computer overlap we were hurriedly stripped of our 7094's to make room for all the 1108 hardware. Too many computers and too little room may have been the cause of this serious disruption.

UNIVAC was either strapped for resources or made bonehead management decisions, when they let out a contract for the design and implementation of their Job Control Language (JCL) to Computer Sciences Corp. What a profound miscalculation! This meant that upon delivery of the system Computer Sciences Corp would disband its work force as was common with contracting software companies. UNIVAC would not have an experienced team to oversee the early implementation of the operating system. I can't imagine how they expected to pull this off. Later JPL woke up and regretfully realized that many software projects were not going to be completed on time. When the fallout from this miscalculation engulfed both the Navy and NASA, an experienced UNIVAC systems engineer could name his salary.

This predicament was actually not that unusual, as new computer systems of that era were notoriously unpredictable. Corporate judgment was widely known to be flawed on most new products. Delivery deadlines were often missed, but this was not going to be just a slight inconvenience. Al knew this and had seen it before. I remember we were walking out to the east parking lot, when Al's words struck me. "An already impossible job was going to be more impossible... how do you program a computer that doesn't work!" My heart rate went up twenty-five percent. The next day I cornered Al for a talk. We quickly realized that we had to get DPTRAJ running and fast. Al added prophetically "when the traffic builds up that computer is going to fail!" He added, "We might have a brief window of three or four weeks, where a semblance of stability may occur." We had to take advantage of this opportunity but it was still a long shot. I was flushed with excitement as it seemed like "mission impossible."

I had always strived for programming simplicity so I figured my subroutines should compile easily although the overall interface with other Links would remain uncertain. The numbers coming out were undoubtedly junk; I was finding new bugs every day with no end in sight. Sensing the sting of battle, day and night we worked. We were the first to start in the morning and the last to leave. JPL slept. Al and I used bootstrap and bailing wire, but we actually got DPTRAJ to run...not reliably...but it ran. We were experimenting and unraveling the mysteries of the new computer. There were no adequate manuals available. No one from UNIVAC was much help in explaining how it functioned. This was about as exciting as it gets. It was interesting to compare the 7094 numbers and the 1108 numbers. We freed up those 32K restrictions to 64K and actually got a four-minute trajectory to Mars. However, the roof was about to fall in.

The next time I remember anything especially memorable was about six months later. No software development had been possible for many months. At a large auditorium meeting there was an impassioned plea by one of Dave Hilt's new young programmers, "Please give us some help! The computers don't work and we have a mission to fly." Near tears, he had expressed all our frustrations. The tension at the meeting was thick and it was hard to tell where it was going. Explosive comments about UNIVAC seemed about to be vented. Then a senior advisor from upper level JPL management stood up and calmly said "Listen, we have had computer problems before and we will solve this one." His brief statement brought calm to most of us. I mark that point as the end of the beginning of the UNIVAC war.

There were many obstacles left, but after that meeting everyone rolled up their sleeves and intensified their work. The crisis slowly diminished. Many years later I was told that there were some electrical resistors soldered into the bleeding Bakelite of the integrated circuits on the 1108. JPL had finally learned to love the 1108. Usually you would scrap a CPU with that sort of serious design flaw. During the crisis I had pulled back into a mathematical cocoon. I was checking out my program without a computer. I got one real computer run per week, and even that was usually suspect.

The Uniscope was the first video monitor available on the desk of a programmer. With a keyboard it theoretically was intended to make cards obsolete. This did put you in competition with dozens of other programmers hoping to get the attention of the CPU at the same time. The stories surrounding the Uniscope are legendary. I spoke with programmers who had waited like a coil of spring steel to press the button that would execute their program. This brought intensity to your desktop. Some determined programmers actually sat for hours waiting for the blinking light to alert them to start their program. Sometimes they waited all day long for nothing.

I decided not to participate in a futile attempt to execute programs on the Uniscope. In fact, I never used a Uniscope and I continued to prefer the old IBM cards, i.e. submit them and let someone else worry about running them. Cards are passive, non-intrusive, and you can even write on them. Carry them in your pocket, they're just the right size. The Uniscope relentlessly demanded your attention like a TV set. As the multiprogramming environment improved some of this early tension relaxed, but "set it and forget it" was not yet a part of our tenuous predicament.

The crisis was not only at JPL; other NASA installations and the Defense Dept were impacted. Virtually all third generation computers were experiencing some of the same serious trouble. The UNIVAC problems went well beyond the operating system. IBM held a patent on card readers. The IBM reader moved the card horizontally across some tiny brushes. The UNIVAC card reader wrapped it around a drum while reading it and then slapped it flat. A deck of cards sometimes looked like a rag after four or five runs. Jams were frequent and violent.

I opted not to sit in front of the Uniscope and the card reader was a dubious alternative. I developed a unique method of visually checking out my coding. Visual debugging means quite simply reading it and analyzing it until every bug is located. I slowed way down and caught many of the hasty mistakes I might have made before the computer snarl started. That is not to say everything was debugged by just "looking at it." However, my new code was almost diagnostic free and this helped a lot. The older code was on tape and I dissected it with the same careful methodology. I felt gratified to get one decent computer run in a week. The real progress was inside my head as I read the erstwhile listing.

I was surprised one day when Mike asked me to speak with the UNIVAC systems group. I was a familiar face in the computer room from the beginning. They were asking me to help them directly with their system. This was a somewhat dubious compliment. For me to work on an operating system required "tying and gagging." It was now clear how serious their desperation had become.

I doubted that I would be much help but I was willing to try. First I wanted to know about the multi CPU situation. Al was suspicious that two CPU's working on the same group of programs simultaneously might be theoretically interesting, but practically disastrous. My attitude was "simpler was better" until things calmed down. I was told the systems engineers had already limited the multi-programming environment to one CPU. They were even briefly operating in batch mode. UNIVAC sadly did not develop the operating system "in house," and the lack of experienced people was apparent. There is no substitute for personnel who have grown up with a system from its inception.

It seemed to me that before you could incorporate some of the spectacular third generation computer innovations you needed to tame the CPU. We couldn't get two runs in a row to execute with the same timeline. I suggested they first obtain a repeatable baseline operation and then gradually introduce the more complex interactions. It was theoretically possible the advanced third generation concepts eventually would handle JPL's anticipated heavy traffic. DPTRAJ was functioning as a hardware diagnostic tool. They said it was far better than the prepared system diagnostics because of its size and complexity. Their plan was to run DPTRAJ until the 1108 malfunctioned and then fix the 1108. This surprising procedure was repeated day after day.

I wanted to know where the CPU was spending its time and how it transitioned from task to task. Without documentation I was trying to discover the strategies used by the system, but the CPU on the 1108 seemed very elusive and unpredictable. The IBM 7094 would allow you to trace sequentially the input-process-output of each program. The UNIVAC, on the other hand, would jump around, calculate here, and calculate there, without a predictable pattern. This is called multi-programming and its creators, Computer Sciences Corp., were not there to help. The documentation seemed obscure or nonexistent, and I was left with the task of discovering its organic structure by watching it.

Dumps, such as we used with the 7094, were not going to help. I was told you could not trap the CPU and obtain the last group of instructions that had executed. It ran your program in tiny slices of time. I was given a fully devoted computer to see if I could unravel what was happening. I used printed clock times in my FORTRAN program to follow the CPU's spurious path. It was surprisingly inconsistent, showing attention gaps of unaccounted time. This was puzzling since there were no other programs executing. I became keenly aware of my lack of systems experience. That left me with a couple of possibilities; maybe there were bookkeeping duties or even timing cycles sandwiched in between program execution. I did not feel I was helping very much and wanted to get back to the DPODP.

UNIVAC's main random storage was called the FASTRAN drum. This was a monstrous device five feet high, ten feet long, and weighing in at two and a half tons. A massive pair of drums was spun at a prodigious speed of 880 revolutions per minute. The calculation of its kinetic energy was daunting. This was the zenith of "mass vs. mass storage." Some estimated that it could have destroyed the Space Flight Operation Facility building if it flew apart. I don't believe it ever functioned reliably for more than a few hours at a time. It was soon replaced with the much more reliable IBM drums that were strapped on. I often wondered why we didn't get an IBM card reader as well.

Looking back at the third generation 1108, the two CPU concept was exciting to think about. We were about to transcend batch processing to multi-programming and beyond, while upgrading the computer brain from one to two. Wow! Picture this: while the CPU is jumping from task to task you have a second CPU somehow competing for your job. Today's high-speed chips make these nano-second machines and exotic sounding processes seem very awkward.

I understand today's modern standards for software management is highly structured. In the 1960s every programmer developed his or her own private technique for debugging, and I wonder how I would have reacted to the modern managed methods. I know structuring helps limit programming disasters, but gone is the freedom to follow your creative whims.

Programming is not for everyone. I know of many practitioners who would be better suited for other work. I was given full authority for programming decisions. My unfettered world was totally free and without interference. I aggressively challenged everyone connected to my work, and without realizing it I accidentally and unintentionally stepped on a few toes.

Informatics instituted briefly a company-wide requirement for weekly progress reports. I had long ago separated myself from them. Even though they paid my salary I rarely had any contact with them. I was never given any technical direction from them, as they were ill equipped to help me in any way. Feeling progress reports were unnecessary, I fired off a couple of terse one-page weekly reports. They must have found their way back to JPL and created quite a stir. The reports not only criticized DPTRAJ, but also criticized other related work. One of my remarks caused Lee Laxdol to order all DPODP programming stopped. He had everyone documenting for three weeks. Skip Newhall was the manager of the DPTRAJ group at that time. You could never disrupt his good nature; however he did mention that rocking the boat did no one any good. The relaxed and open atmosphere of our project needed little regulation, as we were all overachievers. I was deservedly on everyone's list for a few weeks.

The most common technique I developed for locating bugs was to rehearse the program execution over and over in my mind. I began with the input, which led to the starting procedure, derivative box, step-wise integration, dependent and independent triggers. I could visualize the spacecraft, the equations and the coding combined and inseparable. With patience I might slowly page through the program listing; sometimes, however, I found myself just staring at a wall. I could spend hours relentlessly rending the program structure.

In spite of what seemed like a pointless repetitious exercise, gradually a small logical crack and an uncomfortable feeling would emerge. Then it was back to the listing and a new BUG would be found. It is hard to describe the excitement and satisfaction of this work.

It was said that Bobby Fischer was able to examine chess strategies twenty-one moves ahead. I could have used him here.

WEEKEND WALK—Four computer programmers from Pasadena's Jet Propulsion Laboratory, from the left, John Strand, Skip Newhall, Dale Boggs and Richard Brodie (seated), have programmed a trek across the heart of Death Valley, in special heat-resisters. HIKING BEAUTY QUEEN—Kathy Pieper of Fresno will join party of JPL hikers.

Beauty Joins JPL Death Valley Hike

By JIM MARUGG
Staff Writer

A beauty queen from Fresno and four computer programmers from Jet Propulsion Laboratory will hike this weekend across the center of Death Valley.

The hikers believe the 35-mile hike through the very heart of the desert has never been accomplished before.

"The Frenchman who made the hike two years ago took a route near the perimeter of the valley," said Skip Newhall, one of the hikers. "He was never as far from help as we will be. His way was watched by a following car most of the way."

The other JPL hikers are Richard Brodie, Dale Boggs and John Strand.

The beauty queen is Kathy Pieper of Fresno who holds the crown as Miss AAU California.

To make the trip as cool as possible across the blazing valley they will wear cloaks of a towel-like material soaked in ice water, "mainly to keep our heads cool," Newhall explains. "Pockets in the cloaks will be stuffed with ice to keep the cloaks wet."

The trip will begin on the west side at Telescope Park. The first leg of the trip will be all downhill to Shorty's Well at the western edge of the salt flats. Shorty's Well has water but it isn't drinkable.

They'll hike across the salt flats to Badwater, the lowest point in the western hemisphere. There they will be in contact with their support party which will meet them by car. They will spend the night at Badwater.

The next day will be all uphill to Dante's View, an overlook on the east side of the valley. They will be met there by their support party late Sunday.

U. S. Rangers recommended they not hike up to Dante's View because so many hikers have been stranded on the way up.

"There is a lot of shale and the loose ground makes hiking difficult," said Newhall. However, he has made that hike and is confident it can be done again.

The rangers advised against the whole trip, but were adamant about only one thing, that the hikers be met at Badwater by their support party.

"We didn't have to be talked into that idea," Newhall said. "We wouldn't think of taking the trip otherwise."

The hiking party will carry new concentrate, Gator A, developed by the University of Florida, as a supplement drink.

"It has the elements that water doesn't have but that is needed when anyone sweats a lot," Newhall said. "It's a good preventive for dehydration. Lakers use it. A lot of football players use it. It tastes like lemonade.

The four JPL men will drive up late today. They will meet Miss Pieper there. The hike will start Saturday morning. Newhall believes they've taken necessary precautions for a safe trip. But why are they trying such a hazardous, difficult hike in the first place?

"It has been said before but I guess I have to say because it's there," said Newhall. "It's a challenge, particularly after rangers advised against it."

Courtesy: Pasadena Star News

CHAPTER 10

JPL FRIENDS (DEATH VALLEY OR BUST)

Rich Brodie, Skip Newhall, Dale Boggs and I came up with the plan to hike across Death Valley. It began as idle talk and soon grew into a challenge just to see if we could do it. Why not in August when temperatures could soar to a hundred and twenty-four degrees? After all, since we worked together why not have a little hot fun and experiment with the elements? Rich contacted the Pasadena Star News, hoping for a little publicity. Perhaps we could round up some funding for our excursion. It was also Rich's idea to contact Miss California and invite her along.

We trained out at the desert for several weekends before the actual hike. It was necessary to know how we would withstand the extreme temperatures. We found that the most difficult part was when the top of your head gets hot. Unless you could keep it cool with water or ice, you would pass out. Rich seemed to have the most trouble with this and several times had to be rushed to the air-conditioned car that was standing by. He insisted that he would make it when the time came for the hike. Skip and I even tried to jog. It didn't last very long as it totally exhausted us.

As far as clothing went, we decided that terry cloth or something heavy would work best for us. Every five or ten minutes we would pour water over our heads and the heavy cloth would soak up and hold the water. We also brought jugs of solid ice that of course melted into cool water. Skip had managed to order a fairly new product called Gatorade. At that time Gatorade was only available for institutional use, not for sale to the general public.

Our plan was simple: start at a higher level, walk down to the valley floor, cross it, and then climb the opposite mountain. That was the plan!

The day we had been training for finally arrived. The air-conditioned car couldn't make it up the first hill, so it would have to stay around the edge of the valley floor. Rich had developed a clever device for the hike. He had a rubber squeeze tube in his pocket that would pump water out of his pack and up to the top of his pith helmet and then over his head! During the first leg of the journey down the hill, Rich and Miss California were lagging behind. Sure enough, Rich was ready to pass out. Leaning against a rock, he had dislodged his water tube and the hose was spilling it all out on the desert floor! Everyone recognized a real emergency and the car was a great distance away. Out came a mirror and Skip flashed a signal to his wife in the car. "Mayday! Mayday! Please start driving in this direction. We need help!" That was what the signal was supposed to mean. Skip's wife ignored it, thinking he was just testing the mirror flashing system. Finally Skip had to run down the hill to tell her it was a real emergency. They were able to drive the car partway up the very rocky path. Meanwhile, Rich managed to move further down the hill. Since he had no more water in his pack, he started pouring Gatorade on his head and face. Needless to say he was a bit sticky by the time he got to the car. At that point it was decided that it was too dangerous for Rich to go any further. The hike was called off.

This would be a good time to tell you about Rich's fantasy. While on the hike, Rich had taken some eight millimeter film clips with his handy movie camera. He was planning to put together a short clip like you see between double features. He would then take Miss California to the drive-in and surprise her when the short feature came on. Needless to say that didn't happen.

The day after the scheduled hike, the phones at JPL started ringing off their hooks. It seems the Pasadena Star News article made it sound like JPL had something to do with our little hike. Hmmm? Wonder how that happened? Anyway, people were inquiring how the Death Valley test of space suits had gone. Public relations immediately showed up at our office door, asking:

> "Just what are you guys up to now?"

IBM VS. INFORMATICS

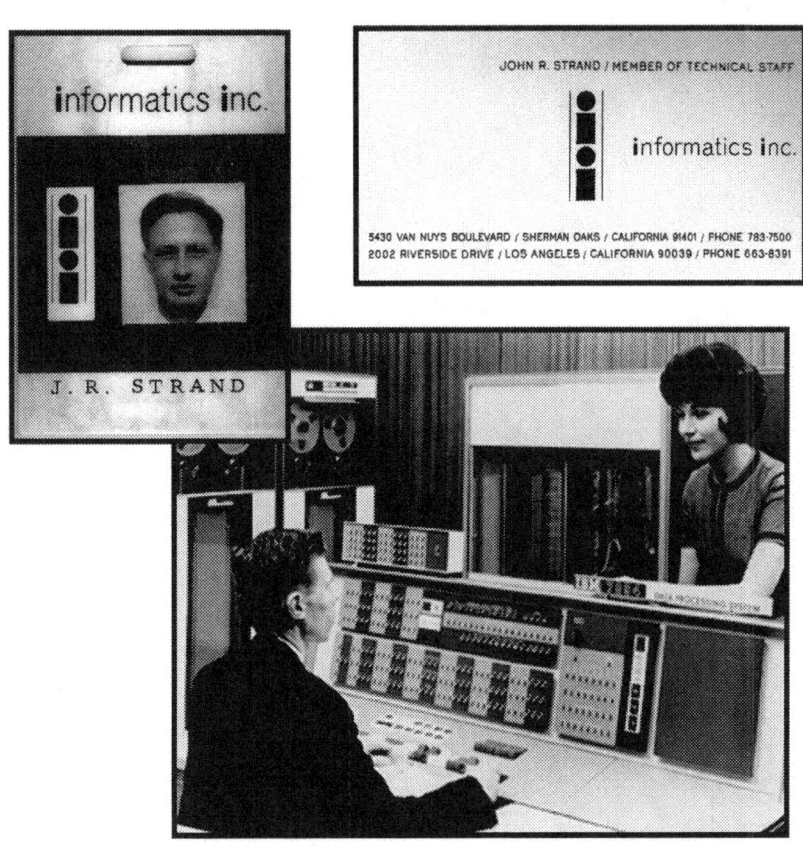

Drum in background was treated as a tape drive by the IBM system.

Courtesy: IBM Archives

CHAPTER 11

IBM VS. INFORMATICS

IBM in the 1960s was a very well run company in full control of the world's computer market. This is similar to Microsoft controlling most of today's operating systems. It treated its employees fairly, but did require strict dress codes: white shirts only, no pin stripes, and recommended suit colors. There were a few unwritten rules--hair length, skirt length, etc.--all of which were intended to create a certain look as well as attitude with its employees. Non-conformity was frowned upon. This was not as bad as Henry Ford's dictums, but went well beyond most corporate standards. Most employees did not resent this because if you played the game you would be well rewarded with promotions and salary.

Informatics was a flyspeck on IBM's windshield. There were only two of us from Informatics at JPL and there were hundreds from IBM. The JPL computer infrastructure was completely controlled by IBM in 1966. IBM Federal Systems supplied contract software personnel. This resembled Informatics' relationship with JPL; however, IBM had nearly twenty contractors with some connection to the DPODP. I was the only Informatics contractor.

I was the fortieth programmer hired by Informatics. Walter Bauer, Dr Rector and a secretary met one evening for coffee and Informatics was born. Eventually there would be three thousand employees working for Informatics in the 1980s. It became the largest contract software company in the world. In 1986, following a hostile takeover, Walter Bauer was removed as CEO.

We didn't feel very big in 1966. IBM had a manager that was a decent sort of person, but seemed a bit catty and looked down somewhat at my position. He fielded the majority of the DPODP team and was very competitive. I was hardly a threat to the IBM limousine.

I had launched myself fully into ODINA, the input Link of DPODP, and used the inclusiveness of the input to springboard myself into the analysis. I wasn't yet cognizant over any analysis, but Mike Warner appreciated my desire to learn as much as I could about the program. The input was massive but did mirror all the features of a large and comprehensive orbit determination program. At some point I realized that much of the input was the a priori set of parameters whose values were to be improved by the correction process. The core storage of the IBM 7094 would not accommodate all the input and a second Link, ODINB, would be required for that purpose. The input Links transferred the user input to I/O files for use by other Links. They also did a small amount of validation to see if there were inconsistencies in the data.

JPL provided its own group of systems engineers for the 7094. They were very helpful and assisted us with our many system difficulties. I became concerned about the parameter updating that would eventually be necessary for my input files. Somehow the systems group located a serious deficiency in the operating system and they shared it with me. On the surface it appeared the inherent sequential nature of input-output files would not allow us to change one parameter without rewriting the thousands of other parameters that follow. This was a combined problem concerning directly addressable, random and sequential computer data handling. IBM provided drums with the physical capabilities of random access but the operating system was designed as if the drums were a tape drive. It was clearly impossible to update a tape record in the middle of a tape without rewriting the entire tape. The drum software mimicked a tape. Business systems were usually "record by record" sequentially. IBM was first and foremost a business-oriented company. Random access was not available on the 7094, since most business applications sequentially processed records.

Remember, anytime you move anything physically, whether it is a card reader, card punch, printer, tape drive or drum, you slow the computing to a snail's pace. Computing or number crunching moves with the speed of electricity. Because one program is run to completion (batch mode), physical motion becomes the limiting factor determining computer speed. The Direct Couple principle of the 7094 helped with this, but it was impossible to update an input parameter without slowing the calculation process. In fact, the input-output would have crippled the program and made it unreasonably inefficient. IBM to the rescue!

Mike discussed this with the IBM manager who said, "No problem, we are IBM and we solve every problem." They were going to use program logic to simulate random storage. JPL's own systems group gave me a completely different picture. They strongly stated this was not possible.

Historically there had been a number of computer software victories that were legendary. Problems advertised as impossible had been solved by small groups of dedicated programmers. The stage was set for a serious confrontation. With the advice and tacit backing of the in-house systems group, it was Informatics versus IBM. David versus Goliath.

I could tell that JPL's systems group was forthright and genuine. IBM Federal Systems looked a bit like window dressing and in my opinion were being paid for more than they were producing. More show than go. I was competing with twenty people who had IBM on their side. Guess who won? (I was always right on computer matters).

In retrospect there was a prevalent attitude in the 1960s of "can do!" The Marines think this way. A few people against overwhelming odds accomplished many surprising computer victories. This was not to be for IBM.

I guess I have to say that this was satisfying. Mike, I think, was pulling for me from the beginning and this was the type of support I appreciated. He released (laid off) the IBM hoards. Never again would we try manpower to solve our problems.

Above: Drum Machine
Right: Keypunch Machine
Below: IBM 7094

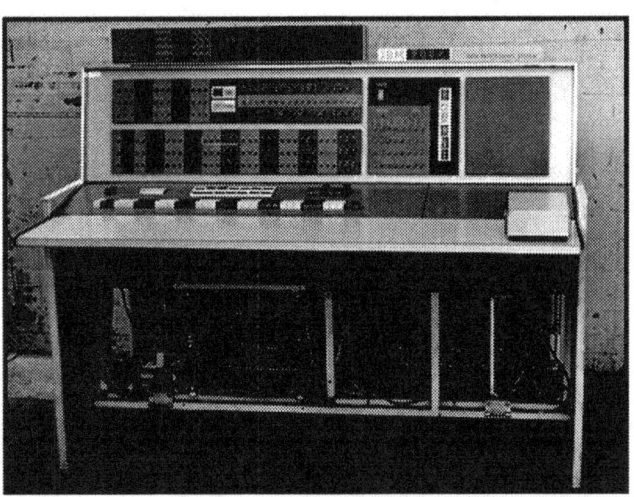

Courtesy: Paul Pierce   www.piercefuller.com

CHAPTER 12

RICH BRODIE, INVENTOR

Rich Brodie put together the early execution of the Link PATH, which was the first numerical integration package inside DPTRAJ. His early work was performed on the IBM 7094 computer. He had worked on removing the diagnostic statements and had managed to kluge up the program to get it to load. Because the overlay was so excessive, to fit it all together required a great deal of manipulation. Rich would spend all day to find two or three extra cells of core to allow the program to load. On his first trial of the program, he reported that it would take longer to execute the program than it would actually take to fly to Mars. This of course, was not going to be acceptable, but at least it was a start. Mike Warner was also happy to report some progress.

Rich believed "the best government was no government." Ayn Rand was his idol. It was Rich's idea to resign from JPL on his twenty-sixth birthday. He claimed he was waiting for that exact day to avoid the Vietnam War draft. JPL was providing a deferment. Rich wanted to go off on his own and experiment with his various ideas and inventions.

After leaving JPL Rich took a part-time job with the Northwood's Inn as a waiter. This allowed him to have his days free to work on his inventions. He claimed with tips he made as much as JPL paid him. Everyone agreed that he had some pretty remarkable ideas up his sleeve. To my knowledge none of those inventions made Rich any money.

As a teenager, the Kent cigarette jingle had caught Rich's ear. It was the one that went "to a smoker it's a Kent." With research he found that these jingles were called Quatrains. So he proceeded to write 101 KENTrains (Example: "to a gold miner it's a lucky strike"). He felt these would be useful for the commercials. Since it was expensive to have a book published, he learned how to do it himself. Rich bound up about thirty copies of 101 KENTrains. Full of self-confidence, he boarded a plane for New York City. At the Kent offices he was directed to the advertising agency down the street. He left a suitcase full of his books with them, expecting to return the next day to a suitcase full of money. Instead he was informed that they were phasing out that line of commercials. On to the next invention...

Rich came up with a plan that he was sure the newspapers would clamor after: an alphabetized index of all proper nouns in the entire paper listed with page numbers. This would allow the readers to locate information they were interested in without searching the entire paper. For example, Rich explained, Mohamed Ali might be mentioned in the sports section. But he might also be found in other parts of the paper because of his celebrity status. The reader could then easily find everything about Mohamed Ali in that day's paper. Besides the proper nouns, he offered a general subject index for advertising. If you were looking to buy a new television you could easily locate all the ads selling televisions. Proper nouns (person, place or thing) are always capitalized. All the print in a newspaper went through a computer. His computer system (by selecting only the capitalized words) would extract and index these nouns automatically. The first word in a sentence is also capitalized, so Rich had to work on a method to distinguish which of these words were proper nouns. He called this program a "thought module."

By hand he began lifting out every proper noun from the entire newspaper and recording their location on an index file. Remember, the day Rich left JPL was his birthday. He picked up a copy of the Los Angeles Times newspaper. That day happened to have the largest advertising edition. After four months of filing words he was exhausted and wishing he had purchased a smaller newspaper on a different day. It could have been worse, what if his birthday had been on Sunday?! Next he hired a printer to typeset all this information. The printer used a typeset very similar to newspaper print. It filled more than one page and was truly impressive.

There is a national award for newspaper innovation and the Times had won it twice. No paper had won it three times. A third win would put them in the award's hall of fame, which had never been accomplished by any paper. Rich was sure if the Times used his "thought module" they were sure to win the award. His presentation included a large picture of the trophy. Rich walked into the Times Building, expecting immediately to see Mr. Otis Chandler, owner of the Times, and that Chandler would welcome him with open arms. It seems he never really got to speak with anyone named Chandler, but the secretary informed him she would leave all of the information with him. Rich returned a few days later. They passed him on to the data processing manager. Rich described the meeting this way: the manager had a very sour look and appeared not to have a clue as to what Rich was proposing. Next he contacted the Examiner, but since everything was based on the Times they weren't very thrilled. Another plan down the drain.

Computer information to assist the average real estate agent intrigued Rich. Real estate agents were not using computers in the 1960s as a selling tool. Rich developed an innovative concept, which he felt would help agents close real estate deals. He began researching information through rarely used occupancy statistics available from the government. For example, if you were interested in purchasing an apartment complex, Rich's computer program would tell you what percentage of units you could expect to be rented for any given month. He spent a considerable amount of money creating a very attractive brochure. Rich hired a private pilot to fly him around so he could photograph rental property from the air for the brochure. He got thirteen thousand cardboard tubes for mailers. Without any thought about testing his idea he plastered the entire real estate world in southern California. Results: five responses and no sales. Art Linkletter's real estate company made several inquiries; however, Rich felt they were only interested in finding out how he did it! By now you are probably realizing that Rich was ahead of his time.

Each of Rich's ideas seemed to have a fatal flaw. Possibly his most farfetched idea was to publish his own generic brand newspaper. He wanted to produce a paper that was totally computer generated. Because the Pasadena Star News had so inaccurately reported our Death Valley trip, Rich began to think "why spend all the effort to gather information, which was so flawed with error?" Anyone reported in a news story is usually surprised to see how distorted even simple facts are. He believed that people would want to read about auto accidents, house fires, robberies, and cats up trees in spite of the fact that the information is grossly inaccurate. It would be easy to write because it would be all fiction. He figured it would be very profitable, as he wasn't going to hire any reporters.

Another on-going project he had was to build a small computer out of surplus transistors. He soldered together single transistors so his computer could perform arithmetic calculations. How complicated (I mean really complicated) it was to connect all these transistors in a fashion that could perform these calculations! I realized it must have taken months of tinkering. I have heard fewer than a dozen people in the world actually understand how grids of transistors perform arithmetic calculations. Rich's plan was to sell this type of computer to fast food restaurants like McDonalds. The employees could simply push on the button for hamburgers, French fries, soda etc. and the computer would organize it, price it and even tell the employee how much change the customer was going to get back. We know that nowadays this is the way it works, but back then Rich was too far ahead of his time. He did have a problem with this home made computer. He would have to grab it and shake it a bit before it would actually add hamburgers to French fries, etc.

While working on his home made computer, he became intrigued by a new electronics development, light-emitting-diodes. Diodes were like microscopic light bulbs, which Rich found very interesting. They could be connected to hearing aid batteries and inserted into jewelry. He designed "laser jewelry" that was a system of tiny mirrors and tiny batteries. The "laser jewelry" was remarkable to look at. When worn as a necklace it looked like a little spot of light repeated right into your chest. It tended to be a little bit large for jewelry, but was a fascinating idea. Last we heard Rich was off to New York.

    Watch out Swank Jewelry!

Global Image of IO (8-27-1999)

Voyager 2 Launch

Courtesy: NASA Archives

CHAPTER 13

DRS. LAWSON & KROGH

There was no doubt Dr. Charles Lawson was the most popular person at JPL. When Dr. Lawson spoke, you just had to listen. He reminds me of my minister at church. He was an example of the distinguished people that worked at JPL. He was president of the most prestigious numerical analysis society in the country. Charles was also a warm and friendly person.

Mike Warner asked me to speak with Chuck. I spent the entire afternoon discussing the numerical integration situation in DPTRAJ. This "package" as it was called, was not going to be a "drop-in and set some flags" type of system. The vast demands placed on this key function were going to make it program dependent. I was accustomed to the generic type which could be treated somewhat like a black box. The tentacles of triggering and the rigors of number crunching were breaking new ground in the art of numerical integration. Ours would be for one and only one purpose: flying to the planets.

I was almost dizzy leaving Chuck's office that first day. I did wonder a bit why he seemed to have such confidence in me. This only inspired my work. Late nights, weekends, it seemed I was always working.

My first project Links, ODINA and ODINB, allowed me to examine all the input parameters. The Links' purpose was simply to make these parameters available to the balance of the program. This opened up to me the structure of the entire mathematical system. One day I carried into Mike's office a three-foot by three-foot piece of cardboard showing the principal algorithms for the entire program. I wanted to know how the pieces of the puzzle were going to work. Mike was duly impressed. He began feeding me more and more work. Those starting a new job should always try to learn as much as they can about their task.

I was now cycling between Dr. Lawson and a weekly meeting with Mike Warner and Ted Moyer. Numerical integration theory became reality. Mike and Ted talked about the "nuts and bolts" of the DPODP and I was the pipeline to Dr. Lawson. Each week the level of sophistication increased. Slowly I came of age. I was now able to pursue my own instincts.

The weekly meetings were over and I became accustomed to seeing one of Ted's impeccable analysis memos on my desk each month. It surprised me one day when Ted showed up at my office. He began asking me fairly penetrating questions about the work. I explained how I was proceeding and about my continuing contacts with Chuck Lawson. His comment was "you seem to be the only one here who knows what they are doing." I was a "made man" now.

Fred Lesh was part of the early celestial mechanics work at JPL. Fred wanted everyone in the Programming group to read a book two hours each day. He felt it was important to stay up with our constantly changing technology. Fred often gave a presentation on some aspect of celestial mechanics. Everyone always paid attention when Fred spoke. He was full of information and always brought his talks to an exciting climax. One of his presentations had to do with simple trigonometry functions. Everyone agreed the standard library functions from IBM were inadequate. I found it very interesting that for small angles you would need at least three different algorithms for sine and cosine. Tiny, teeny, teensy-weensy I guess. The smallest of these three was amazing; you could actually see where the best approximation was the simplest, which was unusual.

The last I saw of Fred, he was doing wind sprints in the JPL Park during a lunch picnic. The JPL Park was located in the Arroyo Seco Wash behind the buildings. This is the same park in which Charles Richter (of Richter scale fame) was found unconscious after he took a walk. Von Karman also launched his rockets here. This old "wash" has quite a bit of history to it. Walk six hundred yards beyond it and you are at the Rose Bowl.

Dr. Fred Krogh came late to the DPODP game, but was critically important. Dr. Lawson, who could recognize genius immediately, hired him away from an aerospace facility. When I first met Fred, he gave me several briefs on numerical methods. I sensed he was going to play an important role. He was a resource for the pure mathematical principles we would need.

The deep mathematical truths, which DPODP was eliciting, would change forever our view of applied mathematics. When you look deeply into "hack and slash" numerical methods you uncover some of the most incredible truths in pure mathematics. (This is the reverse of the rule that applied math lags one hundred and fifty years behind pure math.)

Years later I was looking through some information at Humboldt University and was surprised to see the most comprehensive numerical library functions in the world. Dr. Charles Lawson and Dr. Fred Krogh were named among those responsible for generating them. I heard that Dr. Krogh might have a numerical integration method named after him. I would bet he got some of his ideas from... DPTRAJ.

IBM 7094 Printer

A section of the picture of the voyager tour montage

Courtesy: NASA Archives, and IBM Archives

CHAPTER 14

JOHN EKELUND

John Ekelund was the only other "contract" employee at JPL who contributed significantly to the program besides me. JPL hired us as contract employees because of an agreement with the city of Pasadena. The city limited JPL to five thousand direct employees that were paid and directed by Cal Tech. JPL had literally hired another five thousand from independent companies who were like the regular employees but would have an office somewhere else. Mine was in Los Angeles. John's situation was basically the same.

John resigned from the lab before I did and was very well respected. He completed an important Link called VARY, which solved what were referred to as the variational equations. They were basically partial derivatives relating to the equations of motion and were principally used to enable the program to improve the accuracy of the initial a priori parameters. It was a very complex Link. It was connected to DPTRAJ in a parallel capacity. VARY was slaved to the same step size process I followed. He also relied on my part of DPTRAJ to supply information to his numerical integration. I would pass to him calculated values of the unbalanced forces that were on the probe. This eliminated the need to re-calculate them.

John was on board before I joined the lab and was one of the first programmers on the project. He knew Earl Grossnickle, and like Earl had a lot of technical savvy. John had a tough, unyielding attitude about quality work. We related well on this topic and I considered him one of my sidekicks. One day he was talking about completing a job. He said "I told them I could do it, but there are two basic things I require: 'Pay me and be willing to wait'." With his voice, mannerisms, and wry sense of humor, he reminded me of Johnny Cash. I can remember one day Mike Warner happened to be sweeping out his office and about the only thing John said all day was "you look pretty good behind a broom." He mentioned that we would be old men by the time we saw pictures of Uranus and Neptune (He was wrong).

John Ekelund returned to the lab and is now getting ready to retire. His familiar remark was "When you leave the Lab, no matter who you are, you are blamed for everything." He is probably one of the only programmers left from that early period that is still working with Orbit Determination. I understand that he speaks around the world. John left the lab before the Congressional Award was given for DPTRAJ, and should have been high on the list of the people that received it. VARY is inextricably connected to DPTRAJ, although considered part of the DPODP. Since he resigned from the lab early and perhaps because he was a contractor he didn't receive the award.

I had a small problem beyond my control in receiving recognition as a contractor. There was a move underway to remove all contractors' names from JPL documents. I had written a very deep stack of these and was already named on many of them. Gerd Spier was hired to write the "textbook" document on DPTRAJ. I was overwhelmed with work but still managed to spend a great deal of time helping Gerd. On the front page of the document he gave special recognition to several key JPL employees and an extra special note about my contribution.

Some objections were raised since I was a contractor. Gerd threatened to resign if my name was removed; it stayed on the document.

Today, astrophysics is considered by many as the most respected of all professions. Historically there is evidence that scientists from the 17th and 18th century were measured by their peers, not in their field of endeavor, but by their contribution to celestial mechanics.

John was the first person that really enlightened me about the importance of the work we were doing. This wasn't something easy for me to understand. Gradually from time to time he would mention that we were in a very unique situation at JPL. We often talked about how difficult and virtually impossible the job seemed; yet when you influence physics you touch everyone in the world. He continued to remind me of the significant historical importance of this work. It was John who pointed out special books. One, for example, was the book Brower and Clemens, a book among books in the area of celestial mechanics. It may have been written back in the 1930s but it was one of those special consolidating books that clearly was a hallmark in understanding the mathematics of planetary motion. Hildebrandt wrote about numerical methods, and Brown's theory of lunar motion was similarly important.

John also realized the important contributions of the older school of astrophysicists at JPL, with Earl Grossnickle being one of them. There had been in the very early stages of JPL, after Sputnik, a group of physicists that worked on tracking satellites. They had to dust off some old books. We are reminded that there had been a very quiet period in the early part of the twentieth century where very little was accomplished in the area of celestial mechanics. I believe you can examine history and see that the nineteenth century was far more productive.

Most of these early astrophysicists at JPL were gone by the time I arrived. Their influence, however, was evident. Ekelund pointed out how Earl Grossnickle had a very good appreciation of the direction the DPODP should follow. John and I spent an enormous amount of time discussing numerical integration; he was someone I could always rely on when contemplating coding strategies. Dr. Lawson and I would discuss theory and formulations of the numerical integration, but little about actual "down in the trenches" coding. However, since John depended totally on me for controlling his numerical integration process we shared ideas. I would usually run things past John first for his opinion. He represented a sounding board, someone who could offer opinions and help develop overall scope.

**Surface of Mars
Viking Lander**

Courtesy: NASA Archives

CHAPTER 15

INSTANTANEOUS MANEUVER

All unbalanced forces on the probe influence the probe's motion as determined by the state vector (three components for position, three components for velocity and time). The probe's position vector originates from the integration central body and is considered inertial. The central body is the strongest gravitational influence ordinarily on the probe, i.e. Sun, planet or moon. "Phase changes" are where the integration central body shares equal influence with another strong gravitational source.

A trip to Mars would entail two phase changes: Earth to Sun and Sun to Mars. The probe's motion smoothly transits a phase change. It is our mathematics that creates an abrupt discontinuity. The probe's motion is oblivious to the vector subtraction, which orients the state vector to the new integration central body. The mathematics is very exact as to where this abrupt discontinuity takes place and is of major concern to the integrity of the numerical solution. This was a precise location where the tug was equal between the bodies. I remember Fred Lesh giving a talk on this exact subject. DPTRAJ had the option to hold the central body fixed, but this was for experimentation only.

The preciseness of the exact moment of time the new gravitational source becomes dominant is not important. The concern is with the vector between the competing central bodies, which becomes attached to the probes state vector. The accuracy of this vector translates immediately into the accuracy of the state vector.

The mathematical purpose for a phase change is to reduce the size of the state vector. If a Mars mission retained the Earth as the integration central body, gradually the increasing distance would reduce the accuracy of the state components. With a Double Precision fixed word length, the least significant part would be gradually lost as the most significant part pushes it out.

Any discontinuity in the state vector, besides a phase change, is of concern. This is why the instantaneous maneuver (IM) model was scrutinized so intensely. It is not comparable to a phase change in magnitude, but important because it potentially could introduce uncertain results. Expanding the solar panels, spring separations and explosive bolts are some of the physical events modeled. These are characterized by a brief but abrupt change in acceleration.

A motor burn is an example of a substantial change in acceleration. This strong force has a duration of minutes and is modeled with a polynomial that allows for the burn chamber to heat up and cool down. There is an outstanding reduction in the integration step size, but otherwise no departure from the way other models are handled.

There is no graduation in acceleration with an IM. Rather it is modeled as a small discontinuity in position and velocity. This simplifying approach to what could be a complicated model was justifiable in the early days of the DPODP. There it remained in relative obscurity for years. To terminate briefly the triggering routine, a "delta t" was added to the independent variable (about a microsecond). This prevented confusion with redundant probe ephemeris records having the same time but different state values. This was consistent with a principle in physics, which states that nothing moves instantaneously and also ties the velocity and time to the position.

Eventually the analysis group went through a review of the IM model and its ramifications to the entire DPODP. There were discordant sounds of concern from certain engineers. Issues were raised that challenged the efficacy of the existing model. Eventually a "Tiger Team" was formed to study the problem. This was a committee representing all sides of the issue.

The first issue concerned Initial Conditions also referred to as Injection Conditions. This is the mathematical dawn of the Orbit Determination process. Prior to this there are the violent and awesome events of launch leading to a parking orbit. This is followed by a burn that transcends the escape velocity of the earth. The cruise phase has begun and all the second order differential equations are waiting to be initialized. Without boundary conditions no calculations in the DPODP can begin. We simply need to know a single state vector of the probe at a moment in time. This will represent Houston relinquishing control of the mission to JPL.

Thus begins a period of about twenty minutes where preliminary estimates of the Injection Conditions are made. Dale Boggs wrote a program for this purpose. It was not part of the DPODP that would have required more computer time to arrive at the same results. A quick fit on the meager data available was its purpose. The DPODP would use this estimate plus DSN telemetry data to immediately improve on the accuracy. This is where angle observables are most useful.

There will be no chance for the mission controllers to relax until accurate initial conditions are verified. This is the end of the beginning for determining these conditions. Throughout the balance of the mission the DPODP will improve the accuracy of these most important six numbers. The time that is somewhat arbitrarily set will not change, i.e. improve.

The Injection Conditions are the mathematical gems among all of the parameters. Because the IM changes the state vector, this is tantamount to reestablishing the Injection Conditions. The concern is understandable.

The second issue for the 'Tiger Team" to struggle with concerned mapping. Models that are inactive for part of a trajectory do not require stepwise integration of their variational partials during the inactive phase. The variational state vectors can map these partials over the inactive intervals. This saves computer time, however questions were raised about mapping across an IM. Especially noteworthy was the change in velocity.

The third issue was a concern I had about the triggering routine and the micro second that the stepwise integration would be deactivated. The IM would be a brief step in time and the dependent triggers would not be searched across. What were the odds of a phase change or shadow crossing at that moment? Slim to none, I suppose. Mission planners could keep the independent triggers away from the event. Ted and I discussed these issues and I believe he felt they were not going to impact our accuracy. Ted's intuition was always excellent and months later the "Tiger Team" agreed that no changes were necessary. The transparency of our work had increased within JPL and additional scrutiny was anticipated.

Harry Lass reminded me very much of my Uncle Jelly Smith. Jelly was a very remarkable person to say the least. He traveled around the world in the 1920's because of a twenty dollar bet. He was at King Tut's tomb just after it opened. As a storyteller and hunter he entertained many famous people at his cabin in the woods of Northern Minnesota. The Mayo brothers, Marty Marion, and the St. Louis Browns included. He had a baseball signed by Satchel Paige. At the young age of fourteen while attending school in New Mexico he was given a rifle to help defend the town against Poncho Villa before Black JAck Pershing came to rescue them. A friend willed him an African hunting safari every year for the rest of his life. He frequently lived in native huts. He worked at Carlsbad Caverns in a basket scraping the walls for guano. Then he went to a farm in Santa Rosa, Ca. to work with Luther Burbank

CHAPTER 16

DAVE HILT

Dave Hilt managed the last and most crucial stages of the DPODP segment. The observables, the light time solution, and the differential correction of the a priori parameters were his responsibilities. In place of matrix inversion, the Householder Transformation was an important step in translating the program to a more efficient tool. There was always overwhelming concern that inverting a seventy by seventy covariance matrix would slow the execution time of a computer run. This would have generated astronomical amounts of number crunching.

The off diagonal elements of the covariance matrix usually approached zero and this caused numerical considerations that were somewhat uncertain. The Householder Transformation was theoretically less accurate than matrix inversion. This might promote more iterations of the least square regression, but since the entire convergent process was much faster it saved total computer time.

Dave also added Kalman filtering. It was a simple polynomial that developed criteria for rejecting spurious telemetry data.

Dave came in one morning and said he consumed five pots of coffee and had not slept at all that night. He carried a large stack of papers and announced he had proved Fermat's last formula. You can't possibly imagine the electricity that swept the Lab that morning. At this date, the late 1960s, mathematicians generally accepted the probable truth of the formula first posed in the nineteenth century. Fermat believed he had proven it, but he left no record or any other evidence of his method. It is written that he said his proof was too large for the margin of his newspaper as he rode on a train. So it would be one hundred and fifty years of conjecture. The only accepted proof of the formula was completed in the late twentieth century and would also not fit in the margin of a newspaper

Gauss, when asked why he wasn't working on a proof, replied he "could think of many other unproven formulas just as interesting." Fermat's formula would become the most notorious unproven theorem of our time.

The news about a possible proof by Dave Hilt spread quickly. We were keenly aware of its historic importance. The gathering was moved to a larger room. I remember that Dave was able to write faster than anyone else I knew. So there stood Dave developing copious equations on the blackboard. He wrote non-stop for about two hours. The room was filling up. I was so busy with my work that I had to hop out and re-submit a computer run from time to time. I didn't want to leave the room with all the excitement. Luckily I was sitting there at a very critical moment.

Harry Lass, an executive at JPL, was also a professor and had written several textbooks. It was rumored around JPL that you had better know the answer to your question before you asked Harry! I believe it was Al Joseph who advised me of Harry's truculent nature. They played golf together and both held a two or three handicap. Harry bore a strong resemblance to my Uncle Jelly Smith. Even his combative actions reminded me of Uncle Jelly.

Dave had set the stage for the climax. He is working his way down the stack of papers. Harry is watching very carefully. Suddenly Harry's hand jerks upward. "NO!" Dave had been developing a group of equations. These had new letters that emerged from his derivations.

I taught my pre-algebra students that a letter has a sign in front of it and a sign hidden. The hidden sign is determined by the value of the letter. If the letter represents a negative number the hidden sign is "-". If the letter represents a positive number the hidden sign is "+". With two signs for every letter, a new letter had a new hidden sign, which Dave had not considered. It was this lack of thorough sign testing that had caused Harry to shout "NO!"

Without additional analysis the proof would remain incomplete. Dave had given his all but it would be another thirty years before this work would be completed. I thank Dave for this exciting morning. I wonder what he thinks of the first completed proof, and was he on the right track?

Harry Lass remarked that it was the most direct algebraic attempt at a proof he had ever seen.

CHAPTER 17

ALVA JOSEPH

Alva Joseph is a highly skilled computer-programming expert. He managed the DPTRAJ JPL staff for six months and was a very close friend of mine. I believe Al and I have had more discussions about the main programs in DPTRAJ/ DPODP than any other subject, except possibly the game of golf. At the end of the day after submitting our last computer runs, we often headed for the Rose Bowl golf course only five minutes away. With two full size courses to choose from, there was usually enough time to squeeze in nine holes before dark. I hit a big fade and he hit a low draw that rolled halfway to the cup. Al gave me a lot of good golfing tips and when he moved up to a new set of Wilson Staffs I got his old classics. As we played different parts of the fairway we had many interesting discussions about programming strategies between five iron and putter.

Al had a very strong opinion about program flow, especially the concatenation of the main program structure and the subroutines. Frequently, coding is inadvertently pushed into the main program that he felt rightfully belonged in subroutines. Each FORTRAN program has many subroutines and functions but just one main program. This places special key responsibility on the main program to delegate but not interfere with the nut and bolt coding of the program. The cogent issue is how much of the highest-level program controls belong in the main program.

When the program loading is complete and execution is about to begin, the CPU looks at the first instruction in the main program, which is also referred to as the driver. From then on an orderly step-by-step process established by the programmer will be followed. The scope and layout of the driver often determines the character of the entire program. Philosophies abound about the proper design of a main program, which is often given casual attention during early development. A price is often paid later for inattention to this critical portion.

Al felt it was imperative that the executive control of the program be retained in the driver. An extreme abuse of this principle is a main program that calls two or three subroutines, which control the balance of the program. If this imbalance is allowed these high level subroutines will call subroutines in a cascading sequence. Logical program decisions such as repeating major program segments are then made deep within this framework leading to unnecessary complexity. Al strongly disagreed with this type of structure and pointed out that PATH, the principal Link in DPTRAJ, which integrated the equations of motion, did not have decisive logic in its main program.

Rich Brodie took over PATH following a hiatus in its development. Documentation and analysis had been the primary focus for over a year. Rich was the cognizant programmer in charge of the coding and planned to leave JPL shortly. There would only be four months for performance evaluation. No one had a clue at that time how much work would eventually be required. Fighting core limitations in the 7094 and uncertainties with the appropriate numerical integration, Rich pulled off a miracle just to get the program into execution.

Hurriedly designed, PATH set forth a foreboding of things to come. Rich was a remarkable pioneer who showed all of us how woefully inadequate our second-generation computer was. Months after Rich left I was given the Link PATH, but sensed a disaster in the making and decided not to touch a single line of code until a complete evaluation of the analysis was completed. I first attempted to understand the numerical integration concepts. This was the heart of the program and would establish a foundation for the entire DPODP to be built upon. Dr. Lawson, Mike Warner and Ted Moyer coached me concerning the ongoing analytical modifications. John Ekelund helped me understand the stepwise relationship I had with his Link VARY. There was a great deal to assimilate and I can't remember the first time I was even aware of the main program in PATH. Having none of Al Joseph's experience with main programs and not really paying that much attention to them, I still had a lot to learn.

The analysis was undergoing radical improvements and I was incorporating these into the faulty framework left by Rich. Most of Rich's coding as was expected had bugs and needed to be modified. But his accomplishments in such a short time at JPL were brilliant. Rich had boldly put together a structure that I would follow for several years. I gradually became aware that much of the program controls resided in a subroutine called trig. This is where the events that interrupt the numerical integration are sequenced. High level decisions, such as when to execute a phase change or turn on a motor burn, are decided here. These were collectively called triggers and this was the type of high level logic Al preferred in the main program.

The numerical integration had a system of controls, which like the triggers interrupted the smooth stepwise process in response to error tolerances. Al would have preferred these controls to be in the driver. We both agreed the program had long passed the point of no return. Decisions made by Rich were now cast in cement. Short of a major effort-consuming overhaul, nothing could be done. Still it was interesting to conjecture on what might have been. It took me some time to understand this principle.

Al had learned from experience that programs he called "disasters" started with poor driver logic. He made a living restructuring main programs. I wonder how many millions of dollars NASA could have saved if these ideas had been widely published in the 1960s.

I can't help but remember the time Al was working on a special project called TRAM. It was supposed to generate programs to run on the computer. There was a mistake in the program and it was generating hundreds of runs. Each run would print about three pages and do two or three page ejects. As Al entered the computer room there were twenty printers each printing a few pages followed by the page ejects. It looked like a choreographer had planned it. Imagine, three pages, eject, eject, happening all over the entire room. It was one of the funniest things he had ever seen. The operators had trouble stopping the program from generating these short runs.

If our numerical integration package had been a "black box," the driver would have set some flags and left the nuts and bolts to the subroutines. The triggers, however, should have been monitored in the main program. Which leaves us with an interesting question: could this program logic be moved to the top or would compilation capacities be exceeded? Compilers do not allow unlimited coding. Statement numbers, symbolic names, arguments and subroutine calls are limited during compilation. It is almost impossible to estimate when you are about to exceed some of these limits until you see a diagnostic message displayed. Splitting the subroutines or reformulating to reduce the coding instructions are some of the typical remedies. Could the main program in PATH support this logic? We can only speculate about this question.

IBM 7094

Courtesy: IBM Archives

UNIVAC 1108

Courtesy of Charles Babbage Institute, University of Minnesota, MN

CHAPTER 18

XX NEWHALL, AKA SKIP NEWHALL

Skip Newhall worked on the JPTRAJ monitor that passed data from program to program during mission operations. He may have had some of the characteristics of today's "computer hackers," but was always a force for good. You may think this example of razzle-dazzle computer wizardry would have a place somewhere in the DPODP. This was not the case.

Computers in those early days were rather isolated and viruses had not been invented anyway. Mike Warner told me one day a young and very impressive man was being interviewed by Wiley Bunton. I was to go and see if we could use him somewhere. He was everything I expected, very hyper and very skilled. He could run off staccato computer wisdom like a typewriter. He reminded me of several pool sharks I had played with. My only question was "but how much math does he know?" I believe this may be the case with today's computer hackers: slim math skills. The young man did get hired on at the lab, but not in our department.

Skip was a wily programmer and competent with math. One Christmas he got a computer program to spell out "Merry Christmas" on the tiny lights displayed on the console board of the 7094. Some of these lights were from the accumulator that was at the heart of the CPU. IBM repairmen were the only technicians familiar with these panels. It was truly remarkable to see it slowly move before your eyes and repeat just like the Good Year Blimp.

As a protest against the Vietnam War, Skip legally changed his name to XX Newhall. He was the only XX listed in the JPL phone book. XX, aka Skip, left JPL to get his Ph.D. at Cal Tech. He later returned to JPL and contributed immensely to our understanding of lunar motion.

If I had any idea where Skip was today, I would "one-ring" him. When we were all working together at JPL we could expect to hear our phones ring once. "One ring" and then a hang up. Answer it and you just got a dial tone. This became a bit of a "one-up-manship" game. It didn't take long to figure out it was Skip who started the trend. Skip mostly rang Mike. Mike "one rang" Skip. Then it spread, I got rang, and reprisals were necessary. However by now several players were involved, so you never knew who it was that initiated the "one ring." You could ring everyone or take your phone off the hook. Alva Joseph shared an office with Skip. Alva said Skip's way of dealing with the "one-ring" was to pick up the phone as soon as it started to ring. Then he would bang it back and forth in the trash can and hang it up by slamming it down as hard as he could. One week Mike thought he had a brief reprieve, since Skip was out of the state…no way. Skip, on the other hand was safe; his motel was unknown to Mike.

Skip was the genius behind the Death Valley Trip. Only that trip was nothing compared to what he really wanted to do. Skip wanted to jump off the Golden Gate Bridge with a rope. He had calculated his inertia, figured the distance to the water, analyzed the problem of the considerable weight on the nylon rope, and the acceleration on his body from the great arc he hoped to make. All very scientific, which got him thinking about other semi-circles.

> Skip's Theory: In watching and checking out arcs and his safety when he made the jump, Skip had figured out a near paradox concerning an auto tire. To an observer standing on the road a speeding car's tire tread has zero velocity when it is in contact with the road. The tire tread farthest from the road (top of the tire) has the speed of the car plus extra speed because it is turning. This turns out to be twice the speed of the car. This is a prodigious acceleration in one half turn. (Something for Firestone to think about). Skip could find no such extreme acceleration from the driver's viewpoint. Thus was the paradox.

This theory was consuming conversation around the coffee machine for weeks (Sometimes I wish we could get the old group together, just like the "Seven Samurai" to solve the world's problems). For the jump, the rope would be attached to the bridge and tied around his body. Remember that this was long before bungee jumping came into fashion. Skip was anxious to test his apparatus. Needless to say the Golden Gate Bridge authorities would have nothing to do with him.

Our last adventure was to ride the rocket roller coaster at the San Pedro Pike. This was the last night before it was torn down. I rode once or twice and Skip went the last trip by himself with hands in the air. No one else was left in line, so Skip may have been the last person to ride this historic part of Los Angeles.

Courtesy: IBM Archives, NASA Archives

CHAPTER 19

TED PAVLOVITCH

I have the greatest respect for computers and the marvelous things they accomplish. I consider myself a mathematician with my forte in the numerical area. I am drawn to computers because of the calculations they are capable of. I am not actually interested in learning about them, but because there is no other way to solve mathematical problems, I must contend with their complexities.

Ted Pavlovitch was very aware of computer complexities and the associated frustrations. He referred to these needless annoyances as "wild-man stuff." Ted did not work on the DPODP, but he was a daily source of help and advice to me. We usually agreed on fundamentals and both of us wanted the simplest approach possible to solve our problems. When we watched football together on weekends he totally amazed me with his knowledge of the game. He knew more about the game than the average television announcer did.

Ted and I were able to attend the week-long seminar on the new UNIVAC 1108 Exec 8 operating system. He was the hardest working programmer at JPL. I looked forward to attending the seminar with him, as he had a fine sense of humor. He was often making light of computer problems.

The early evolution of computers had been a struggle to lift things out of the binary (on/off) electrical state. Assembly language that is now the choice of last resort was in the 1950s a great leap ahead. Gradually the early symbolic languages became more user friendly. FORTRAN (formula translation) meant scientists could in a short time write reasonably complex programs using a text that they were familiar with. Business programming witnessed a similar process evolving into COBOL.

FORTRAN had small differences within each manufacturer's compiler. The control statements that were external system requirements were another story. IBM 7094 was the simplest, just two cards. IBJOB and IBFTC were all that were necessary to execute a FORTRAN program.

Ted and I expected the new "third generation" computers to continue the trend of simpler and easier usage. The less involvement with the system the better, freeing up our time so we could concentrate on our scientific problems.

This was not to be. In fact, the "third generation" system commands (JCL) resembled its own symbolic language that we as FORTRAN programmers found very frustrating. Ted believed the more system commands necessary to execute a simple FORTRAN program, the less desirable the entire computer would be. This simple test would tell us what we were in for with the new 1108.

At the UNIVAC 1108 seminar, Ted asked some very penetrating questions. The instructor was filling the blackboard with nonsense syllables (acronyms) that we were going to need to execute our FORTRAN. I had to hold myself back, as Ted could be very funny. I heard him say under his breath, "wild-man stuff," numerous times. After about two hours, we all took a break. Ted analyzed what was going on and I wished I had some pain pills. I left at lunch with cramps and never went back. He stayed for the rest of that day, but did not bother to return the next day.

Ted concluded there must have been two separate groups creating the operating system. They had not been in close communication. There may have been reasons such as schedule deadlines, but at some point a re-design should have occurred. Instead, compromises between the two groups created a profusion of mandatory control commands. I could sense a lack of continuity and I believe the instructor agreed somewhat with this assessment.

The impact of all of this started to disturb me. I became lost as to what was going on. The humor had subsided and I wondered if I could squeeze this all into my brain. I have always had a funny notion that there is only room for so much in my head. It was already full with DPODP and IBM 7094 stuff.

The internal arithmetic of the 1108 was different. The 7094 used a sign bit and "two's complement" arithmetic. Instead of a sign bit for each number the UNIVAC would now use a "one's complement" arithmetic that had a surprising representation of "minus zero." This was an era when every few years you could expect to learn a new system. I felt foreboding and didn't know if I could handle a new operating system. I did eventually develop a memory problem, but not in the computer or math area. Today I have a real problem remembering people's names.

The emerging features of third generation computers had spawned a new job description. There were going to be programmers whose sole principal language would be the operating system. They would become indispensable, because people like me didn't have the time for all this "wild-man stuff."

Gradually computer operating systems were to become more versatile and not obsolete every few years. I tried to learn as little as possible about them, saving my brain storage area for more important things. I still consider computer manuals from that period the most boring and inaccurate literature in history. Bill Gates obviously took interest and realized a buck was to be made simplifying operating systems.

About this time JPL had uncovered an inconsistency between segments of a computer system used by the Voyager mission. The problem involved two groups, about fifty engineers in each group. One group had assumed a right-hand rule in rotating the spacecraft and the other assumed a left-hand rule. Example: A wood screw if turned clockwise will tighten, but a propane tank connector if turned clockwise will loosen. This illustrates somewhat the right- and left-hand rule.

When the discrepancy was discovered, a meeting involving all one hundred engineers was scheduled to decide on which rule to use. During the hour-long meeting both sides argued their point and left the meeting assuming that the "power of their presence" had convinced the other group. But, no one stood up and asked "which way will it be?" It was much later during the mission that a command was sent to the spacecraft and it rotated the wrong way. It was at this point they realized that they had not settled the question, "Right or Left?"!!

Aside from the 1108 Exec 8 system, Ted had a principle that also seemed to me to be broadly applicable. He on numerous occasions mentioned the rule of one-third. It went this way: "one-third" of the employees are producers, "one-third" actually detract from the work and "one-third" neither add nor detract from the work. When I asked him how anything is accomplished he said it was because the one third that are producers are the more tenacious. I believe he was correct, because later I saw a tape drive dropped carelessly off the end of a delivery truck -- cost for damages one hundred thousand dollars.

Ted was very prophetic when he frequently remarked "The Apollo astronauts better bring more than rocks back with them to justify the enormous cost of project Apollo." (It is notable that after thirty-five years we are now beginning to discuss a return to the moon.)

Charts connected to software progress can be very misleading. Mike Warner with some difficulty designed a large poster size illustration of the progress on DPODP. It was a little like those thermometer posters for charity funds. All the Links were listed to the left and then documentation, flowcharting, coding, execution and checkout were listed across the top. This got us all pumped up to see whose red showed up first. I had several Links and may have won the race.

The reality of scheduling software development in those days was uncertain and imprecise to say the least. There were many examples of software development requiring ten times the estimated time and money.

Gradually everyone got red ink beside his or her Link (except for REGRES). Mike finally took down the chart. The reality was that checkout had a sinister nefarious connotation. It was only in the eye of the beholder and the chart did not actually recognize the Link interface that was going to bog us all down. How much check out is enough? This was a very imperfect science in 1967. I would say that the chart represented ten to twenty percent of the work that would eventually be required. Our completion schedule would be slipped for years and a new, smaller energetic group of programmers would emerge.

Smaller was better since fewer memos were required. I was assigned the equations of motion that would consume me for the next three years. I can assure you this was not a job for "prima donnas." There were reverberations from the analysis group about the impossibility of this program. Mike never doubted we could do it and a new force would enter the picture. This was a light... no; it was a spark... no, not either; this was a "blowtorch" by the name of Al Khatib.

PATHWAYS TO THE PLANETS

CHAPTER 20

AHMAD KHATIB

I remember our first meeting. Ahmad (Al) Khatib was full of enthusiasm and unlimited energy. This was going to bring a new dimension to DPTRAJ. He was the cognizant engineer and we were all going to a new level of excitement. Al always gave me extra incentive to work harder.

Mike was somewhat apologetic, suggesting there could be difficulty working with Al because of his intensity. This didn't concern me a bit. I think I was fairly easy to get along with and high-energy types usually don't affect me. Al was as unique a personality as I have ever met. We were friends from the first day. I knew that here was someone I could work with.

Al grew up among the royal family of Lebanon. His life as a young man was very interesting. Palace life could be challenging for a young boy. Adolescence was especially difficult for him, as dating was actually dangerous. His behavior was held to a high moral standard. Al attended college in the United States. The first person in the USA to attempt a field goal "soccer style" in a college football game was Ahmad Khatib.

Al had a friendly sort of contentiousness. He was very competitive and argumentative. Often we crossed swords over technical details. But, this was all part of Al's unwavering determination to produce the finest program possible.

Some days he seemed like Quantrill "burning and sacking" on a cavalry mission behind enemy lines. I knew when I saw him coming in my direction to just put down my pencil and the battle would be joined.

On the surface Al's memos seemed a bit hurried. Steps in the analysis could be missing, with trails of logic not fully investigated. This was all part of our working relationship. As we settled into a routine, he would bring a new piece of analysis for me to look at. I would locate a detail he had overlooked and he would go back to his office often shaking his head.

Usually all I needed to ask was if he had searched all contiguous quadrants for angle inconsistencies. Quadrants are the easiest place to poke a hole in engineering analysis. Generally the initial design is formulated in the first quadrant. However, angles have a way of translating to other quadrants. Sometimes just the vicinity of the coordinate axis is enough to destroy the most carefully thought out analysis.

This was how we got the job done. Sometimes it wasn't pretty, but it was effective. I am sure group supervisor Fran Sturms was in the wings helping and guiding all along. I believe Fran used Al Khatib like a battering ram on the 315 group. At times I am sure we needed it.

Al was in the analysis group while I was in the programming group. When we were at the blackboard for several hours, there was no distinction between us, each helping the other to solve the problem.

Al always provided an umbrella of protection for me when I needed money for computer time. I could spend $10,000 ($25,000 by today's standards) in an afternoon. Day after day, he never questioned me. He never suggested I somehow cut back on the time, always to the attack. These were especially productive times. I was growing in my abilities as an applied mathematician and Al was the catalyst. The numerical integration was for a while set aside, as newer models were introduced into the equations of motion. The issue of an improved solar pressure model was going to influence our work and briefly shake our friendship.

Al was in the Ph.D. program at UCLA. One day he came in and suggested we replace our XYZ "Newtonian" coordinate system with an exotic relativistic system. This new system would inherently compensate for relativity. Ted Moyer's relativity model was described as an unbalanced force or a correction to the Newtonian reference system. This was very much like Al; he was a risk taker, never afraid to challenge the "status quo."

Coordinate systems should be seen and not heard. They connect us to the mathematics that describes our world. Relativity is a geometry which best explains our universe and is particularly concerned with aspects of our physical world not influenced by coordinate systems. Newton used the XYZ system because his world was three dimensions. He had not seen how time and the speed of light could influence the space around him. There are coordinate systems, which incorporate relativity inherently from the ground up. Al's new system, if used, would supercede Ted's model.

I think and live and understand things as Newton did. This new approach would have stressed my understanding of the physical world and required changes to the models. Gravity may have been plausible with the relativistic coordinates, but attitude control, drag, solar pressure and motor burns were conceptually most simply described in the good old XYZ Newtonian frame. What would they be like in a relativistic world? All the partial derivatives would change for sure. It was impossibly impractical to reformulate at this late a time. But it made for a very interesting afternoon.

As part of Al's Ph.D. program, he was working with a world expert in celestial mechanics at UCLA. The Voyager project had just been conceived. A couple of analysts came by to check our solar pressure, attitude control and deep space atmospheric drag model. It sounded as if there was a fifty-fifty chance we would need to upgrade them. When in doubt, upgrade. Al was asked to improve the accuracy of these models. He decided that a detailed representation of the surfaces of the vehicle should be used. This is a little like an airplane where the wings produce drag and the fuselage produces drag and the stabilizers produce drag. Al allowed for up to eighty surfaces.

This was testing my ability as an analyst. Al was allowing me to participate in this most exciting work. The world of physics would not be the same after we were done. This was deep space, so no boundary layer pressures accumulated like aircraft experience with sonic waves. The mean-free path of the particles, which were cosmic rays, protons or otherwise, was longer than the spacecraft. Here was a possible breakthrough in our understanding of small forces on the spacecraft. Al was excited: his mentor at UCLA gave him a compliment for a possible important breakthrough. Now he must test the waters in my cynical world. I was used to participating in his first couple of attempts at a program modification. As a programmer, usually you agree on certain working requirements and if you hope to function effectively you don't allow modifications after a certain point. We had long ago thrown this theory out the window.

Al came to me with the new program modification in hand. I asked for some references. He stormed out. The next morning he arrived with a "burr under his saddle" and eleven books. My mood was also a bit gnarly. As he started dropping books all over my desk my attitude escalated. Al had a very short fuse that went off in a routine manner. My fuse was long and slow. It finally detonated. I started saying things to Al that I would regret the rest of my life. Later Mike managed to squeeze in between. To defuse me, he suggested I be awarded the Croix De Guerre (WWI French flying medal) for being able to put up with al in the first place. Khatib announced as he left my office that he was going to talk to Pickering and get me fired. Pretty high drama. What a blow-up! I apologized to Al the next time I saw him.

The stature of Al Khatib was evident as he continued working with me as if nothing had happened. This indicated his high level of professional integrity. Years later I had occasion to visit with him. I stayed at his house and his gracious wife fixed me a true Mediterranean meal. But I will always carry with me those few careless words I spoke thirty years ago.

It was Al who told me about the Mercury-Venus mission that was bid just like a used car, ninety-eight million dollars! Just before I left JPL, Al started working on FASTRAJ. It was intended to be a simpler faster DPTRAJ. Logic that was intended for the differential correction process was eliminated or streamlined. I tried to understand how useful it would be to speed up trajectories at the expense of accuracy. I had a feeling the difference between DPTRAJ with its standard options suppressed and FASTRAJ would be negligible... I have got to ask Al someday how it turned out. Al told me he was supplying the early shuttle with portions of its navigation system. I wonder if any of the DPODP was actually used?

Decades later an engineer told me a story. "I had a small group and Khatib came over one day and wanted me to write something that didn't have a useful purpose or didn't make sense. I went to my boss, Fred Lesh, and told him about what Khatib wanted. I was hoping that Fred would say 'tell Khatib to get lost'. Fred had a better idea as usual. He told me to have everyone in the group write a page on the subject that Khatib requested and then have the secretary type what everyone wrote and give it to Khatib. We did no editing or proof reading. A couple of days later Khatib came over and I gave him the document. It was not worth the paper it was written on. Khatib went away and never said anything about the document. He must have thought we were nuts."

**Voyager spacecraft during vibration testing
(3-25-1977)**

Courtesy: NASA Archives

CHAPTER 21

TED MOYER

Ted Moyer approached me one day with an interesting problem concerning his relativity model. Newtonian gravity from the strongest nearby body referred to as the integration central body had been long studied. Einstein was about to enter the world of orbit determination. Newtonian spheres relevant to the integration central body had been a part of DPTRAJ from the beginning. Ted had concluded that since Relativity was a correction to the Newtonian field we should use relativistic spheres of influence. In other words, turn on and off these tiny planetary contributions. If only these additional spheres of influence were based on the same Newtonian gravitation.

Alas, this was not to be. Newton and Einstein were to be joined in a compromise that left Einstein's participation viewed as a small correction to Newton. Each of them would potentially have different spheres of influence. There now could be more dependent triggers and more integration re-starts. I had to think this one through. The sun's relativistic contribution is actually stronger at the surface of the earth than the earth's contribution. This was interesting but not particularly relevant to my programming problem.

These relativistic considerations were not for the purpose of protecting the integrity of the state vector like the regular Newtonian phase change. The problem was all planets were normally contributing Newtonian attraction. If all planets were to contribute in a relativistic way there would be an incomprehensibly small contribution from, say, Pluto while traveling near Saturn. Pluto could be on the other side of the Solar system and hardly significant relativistically. Now adding tiny (but arguably correct forces) did not seem to me to be a problem. I suggested we just turn everything on all the time. This would save restarts and also the dependent trigger monitoring I hoped to avoid.

Now it was Ted's turn to think. He appreciated the difficulty in the trigger overhaul. He felt the sun's contribution could be left active for all trajectories, but had a problem with the tiny virtually insignificant planetary contributions. If you take a tiny piece of insignificance and add it trillions of times, could that create something significant? At this point I trusted Ted, and the decision was to have options to selectively turn on or off planets' contributions during the entire trajectory. This would be part of trajectory planning and not an automatic process. I was relieved that I did not need to significantly alter the triggers. I just tested input flags to determine which planets contributed relativity corrections and these were calculated for the entire trajectory.

I had some other memorable meetings with Ted. When I managed to locate some paper work from orbit programs in other NASA installations, I did a bit of investigating. They were woefully inadequate compared to DPTRAJ, but I did find in one of them an unbalanced force we had not investigated. When a satellite orbits a planet with a magnetic field, any electric current, man made or otherwise, creates a small force on the vehicle. I asked Ted, "Why aren't we modeling this?" Ted, whose analysis was always impeccable, quipped back, "Poor analysis." I broke into sidesplitting laughter.

At still another meeting with Ted, I had come up with a possible disagreement between conventional Newtonian instantaneous F=MA and Einstein's real principle of relativity. The principle isn't actually that nothing travels faster than the speed of light c (which does follow) but no message can be sent faster than c. This means that no one can waggle a massive object and have another object respond instantaneously; in other words, this object will experience a change in velocity, sooner than it takes light to traverse the distance between the objects. This appeared to be a departure from the inverse square law, which assumes a spontaneous action between attracting bodies. You must allow an interval of time for the gravitational forces to react. This would be the time it takes light to travel between them.

I asked Ted about this. He believed his relativistic $1/c^2$ terms were correct. I did, however, have a point and he said that he would think about it. Shortly thereafter, while standing in the cafeteria line, I heard someone discussing just this subject. I knew then that Ted was working on it. I never followed up on why this was not significant. Non-symmetrical matter such as oblateness, along with direct planetary gravity was seemingly not going to be correctly modeled.

Technician checks soil samples on Viking Lander (5-20-1971)

Courtesy: NASA Archives

PATHWAYS TO THE PLANETS

CHAPTER 22

FRAN STURMS

Fran Sturms was a very unassuming person. His status as an analyst at JPL was legendary. He came to me one day and asked if I wanted to work for him in the analysis group. This was a surprise to me, as I did not at that time feel capable of handling original analysis. He assured me I could do it. I wasn't sure how my contractual agreement with Informatics could be handled. Mike had tried twice to arrange to hire me directly. Informatics always insisted on a six-month layoff before I was allowed to work for JPL. Informatics was collecting triple my salary for my services and I had almost no involvement with them at all. Fran was willing to try again.

The more important issue to me was how this would impact DPODP. He was well informed about my ability and I trusted his judgment. I weighed the consequences and the advantages. I was entering a very advanced state of program checkout.

In my fanciful role as programmer I would examine every numerical nut and bolt. I would take them off and measure them, examine them with calipers and polish each washer over and over. But to design their location, that was for the analysis group. I was envious of working on original analysis. It was a dream, but with the likes of Sturms and Moyer, it would be like trying to play guitar with Paul McCartney. Some have remarked that his skill was so brilliant that playing with him nearly destroyed their musical abilities.

My own original work concerned the connection between the analysis and the computer. I remember considering just one equation. It was a generalized acceleration on the probe. I tried to determine how many dozen subroutines its formal analytical representation influenced. The complexities of this program were truly remarkable. Sometimes the cleanest engineering would end up dissipated and undistinguished in the morass of the coding. I believe the analysis group always had respect for the complexities of the DPODP. It was evident we understood their work far better than they understood ours. For programming to proceed it was necessary for me to master the engineering equations and occasionally suggest changes for them. But my real forte was grinding them up and sprinkling them into the program.

There are five hundred equations representing the basic analysis. These are not coded routinely from beginning to end. Like the eight notes on a scale represent all music, these five hundred equations must be massaged by the program logic to produce the correct rhythm. Our synchronization sounded dissonant at this early date. I was too far along the road to stop and work on the original analysis. I told Fran I was more valuable working where I was. If I began to see improvement in our program I would re-evaluate. I had begun to feel no one else could do it. He felt satisfied with my explanation and I believe he realized I was exactly where I could be of the most value. We went over POST, the Link that prints the results of a trajectory case. Karen Kelly wrote POST and was now assigned to another group. There were a couple of options not clearly specified in the original design. We decided to skip them.

Fran and I had a very warm friendship. Fran gave me the finest compliment. I heard it second hand, several years after I left JPL. While he was discussing the success of the Voyager he recalled my work and said "It was like John Strand was up there flying it."

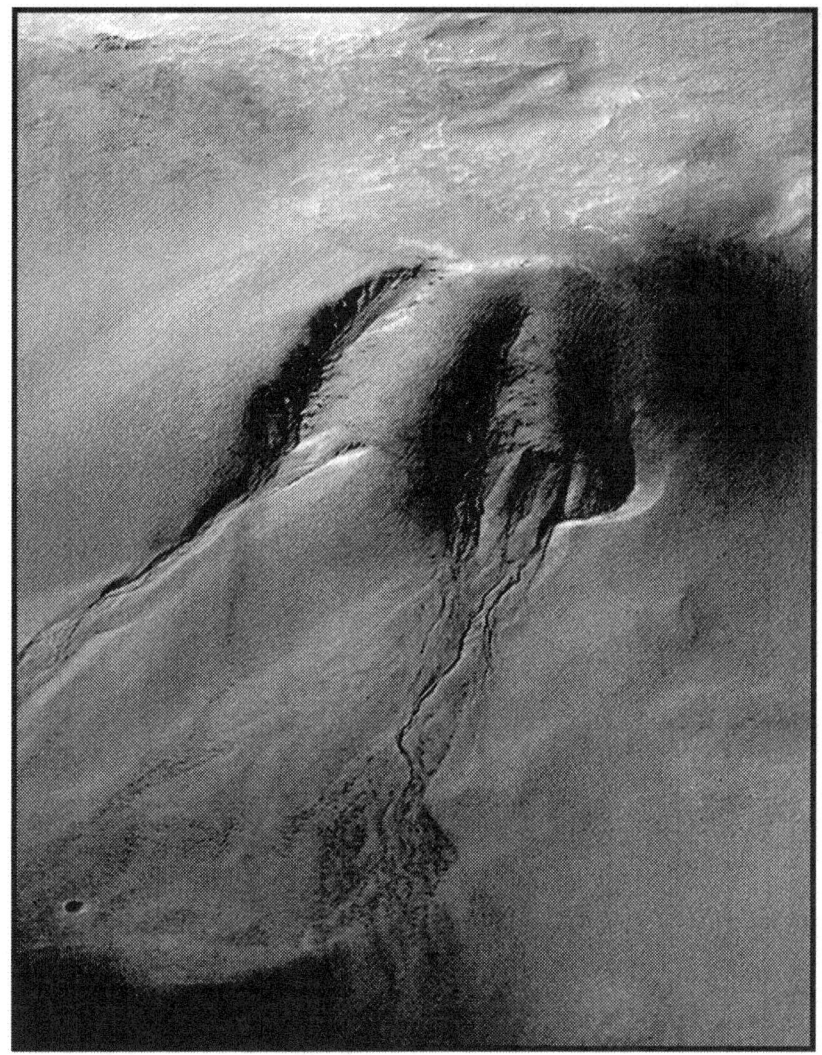

Evidence for recent liquid water on Mars

Courtesy: NASA Archives

CHAPTER 23

DAN ALDERSON

When I think of all the people working at the JPL lab, I think of Dan Alderson as the most complex. He graduated from Cal Tech with at least a 180 IQ. Dan, in his thirties, was still living at home with his mother. He did have two trailers in the backyard, but they were there just to store all of his science fiction books.

Dan and Ted Moyer were the only analysts I knew who understood General Relativity. Dan and I had long conversations about computer languages. At one point he had a favorite language, PL/1, that he felt might replace FORTRAN. He would thoroughly research any topic. This was great for me, because I could then "pick his brain" for the information I needed. I didn't have to waste time reading everything about it myself. This was a little like C-SPAN non-fiction books on the weekend.

There was one facet of Dan's character that demonstrated how complex he was. On a fairly regular basis he would break down and cry. Small disappointments, which we all had, were like mountains to Dan. Very trivial situations would cause him to cry. He sometimes used this tactic if he missed a deadline. This was known by all and everyone just went about their business and ignored it.

When Alva Joseph was brought in to manage part of the programming group, he held a very relaxed initial meeting. He carefully brought up the conditions of our offices. Apparently someone higher up wanted our desks a bit neater looking. Massive computer programs create massive piles of listings. Core dumps were our "meat and potatoes" in the late 1960s. Joe Witt had a way of balancing little piles of IBM cards on every flat surface within six feet of his chair. An earthquake would have set him back a month! I did realize this was part of Joe's method and his work was always excellent (Although his pipe did generate occasional criticism).

Sitting under a sign that read "A CLEAN DESK INDICATES A SICK MIND," Al Joseph continued to explain that he knew that some mess means progress. But, possibly just for show, we would need to make some effort to straighten up for a bit.

Dan Alderson had possibly the most "stuff" on his desk. He asked Al about a few of the papers that he worked with, explaining that they were necessary. "OK" Al said, "If you have some papers that are very important, go ahead and microfilm them."

The graphics of computers was nothing like today. To get a "fast" picture of something out of a computer, you had to use a very expensive process. This process produced microfilm that could be projected on a screen or looked at through a viewer. It cost twenty-five dollars a picture, and that was in 1960s dollars. The meeting was adjourned.

One week later Dan had mastered the microfilm process. Dan proceeded to microfilm absolutely everything he had on his desk. It was brilliant. His desk was neat! All you could see on his desk was a little tiny roll of film. The problem was, at twenty-five dollars a picture, it may have cost JPL twenty thousand dollars! And you know, I think this was the only time I ever saw Al a little unsteady. He had a talk with Dan. And then Dan cried.

Life at JPL went on.

CHAPTER 24

PH.D D D'S

This is a delicate issue to talk about. I feel much like a news reporter covering a story for which the reporter has no position.... "Just the facts, Ma'am."

I can not believe it was by accident that so many key individuals were drawn to JPL at that point in time (1966). JPL was always the West Coast anchor for science and computers, with Cal-Tech the academic connection.

Wiley Bunton was the personnel manager who arranged for my interview. Besides Wiley there must have been some very competent people making decisions on whom to hire for the DPODP. My status as a contractor left me out of personnel matters. Most of the staff were hired about the same time that I was hired. Others were already on JPL projects and were transferred to the DPODP.

Among the programmers, Mike Warner had the only prior orbit determination experience (SPODP). He was the technical and personnel manager but actually did none of the coding. The analysis group, Section 392, was really the central organizing force for the work at the beginning; several of them were already battle tested on previous missions.

As the staff was reduced in size it occurred to me there hadn't been a single Ph.D., either programmer or analyst, working on the DPODP. Dr. Lawson, an eminent numerical analyst, was the only exception. His role was vital to the numerical integration as well as other pure mathematical functions. However, he did not seem to be in direct management linkage, but would step in when his area of expertise was required. Later another Ph.D. I will call Dr. Abbot was hired by Mike and assisted me in getting a grip on some basic numerical strategies. He was not a programmer but did help focus my attention to the Adam's type predictor corrector method we were using. He was most helpful creating a research environment for the integration package.

Before Dr. Abbot arrived it occurred to me we had no Ph.D.'s working directly on the project. At our weekly meeting with Ted and Mike, I asked the question: "With Ph.D.s all over JPL and Cal Tech, why aren't there any working with us?" The answer to this day still astonishes me.

To give some background as to why this was, we can start with TRW. I know nothing at all about their serious attempt to unify orbit determination. I do know they used forty Ph.D.s and failed. That must have been a circus indeed. Funding was no problem in the 1960s. Remember at North American the guideline for solving Apollo problems was "if two people couldn't do the job hire ten." If money was no object then why not hire some Ph.D.s.

Now as the years have gone by, in retrospect I will attempt to explain Mike's answer in a better light. There are clearly some abuses, which come from bureaucratic nonsense. A Ph.D. may be promoted instead of the more qualified non-Ph.D. I went back five years later to take the last couple of courses for my Master's degree and was told I would need to repeat all the courses and their pre-requisites. Politics and degrees are unfortunately linked. A form of hazing is still an academic phenomenon.

Back to Mike's answer. Mike said that in order to earn a Ph.D. you had to do some accommodation or compromise in your studies. To "play the game" for the institution was tantamount to untruthfulness. At JPL if we were to accomplish our extraordinarily difficult task, we would need first and foremost truthfulness from all.

Mike placed truthfulness above everything else. If things were going bad he needed to know. He wanted people who had not experienced the politics of higher learning. I can understand part of this. We also needed to train personnel so they would do the job the way we wanted. By the time you reach Ph.D. levels you have established certain prejudices toward traditional science.

Einstein, when questioned about how he accomplished his remarkable theory, replied "I questioned an axiom." Here was one Ph.D. who was not handicapped by tradition. There are, of course, many, many more.

Venus

CHAPTER 25

LET'S CALL HIM DR. ABBOT

Dr. Abbot was the only exception to Mike's practice of not hiring Ph.D.s for our work. He was just out of college, young, energetic and brilliant. He had the looks of Rock Hudson and naturally was the focus of the young ladies at JPL.

Abbot quickly recognized the importance of our work and he enthusiastically began to question me about it. I do not believe it was Mike's intention for Dr. Abbot to fully participate in the programming since Abbot never showed any inclination towards coding. As far as I know he did not write a single line of FORTRAN, but seemed to periodically work in other areas at JPL. His skill with mathematics was evident and he impressed others by solving some important analytic problems outside the DPODP.

He must have used Eigan values on his doctorial thesis, since he constantly attempted to insert them into the DPODP. Shortly after Abbot arrived he managed to formally arrange with Mike to spend two hours a day with me. These "tutoring" sessions usually went beyond two hours of my time. Abbot learned quickly and had an unquenchable penchant for those Eigan values. After each area we covered I could expect Abbot to bring up Eigan values and each time his attempt to use them was defeated.

I found these clever little mathematical curiosities interesting, but unless they were part of my work I gave little attention to them. Abbot may have found a home for them in other parts of the lab. I could always anticipate when Abbot was about to bring up Eigan values from the gleam in his eyes. This was usually toward the end of our often-productive sessions. While explaining DPTRAJ to Abbot, I found I was getting ideas about researching the numerical integration. At this time the analysis group had gone well beyond our programming and we were constantly in catch-up mode. The analysis was producing beautiful second order differential equations. It was our difficult programming job to solve them.

Abbot began to push me into exploring new techniques of numerical integration. This I tried to find time for, but the heavy load of my other work gave me some hard choices. Everything would eventually need to be completed in the end anyway, so it was a matter of priorities.

The blue print for the numerical integration had originated from Dr. Lawson. To implement these concepts would require advancing the state of the art of numerical integration. There had been some hand waving at these ideas from the beginning. Rich Brodie left patterns of this in his original work on the Link PATH. I steered clear of this in the beginning. Reverse integration, for example, had never been tested. I knew where these options were in the coding. This was very exciting work, but there were other areas that needed my attention. So I attempted to maintain a balance.

Abbot expected more of my time, and he grew a bit frustrated with the pace of things. DPTRAJ was my baby now, and I had to mix Abbot's frustrations with my own sense of priorities. Mike had long ago given me full authority. When Mike and I met I usually had a list of difficulties to discuss. At times they seemed insurmountable, but by the end of our meetings, Mike always had me pumped up again. He would say "I know you can do it," and that's all it took. Mike never had to stop by and evaluate how I was doing, because he knew I would corner him when I was ready. Abbot, on the other hand, would regularly ask me about the status of things he was interested in. By exerting pressure, he did produce results, but he still seemed to know his bounds.

Looking back on our relationship he must have felt powerless, knowing he could never handle any of the coding himself. Somewhere inside him rancor was building, frustration leading to resentment and eventually jealousy. He was outwardly showing very little of this; he was, as it turned out, a very stealthy person. He did unknowingly demonstrate Mike's Ph.D. principle. I probably shouldn't blame it on the Ph.D., but among all those who worked on the DPODP, he was the only person to exhibit dishonesty, Ph.D. or not.

The first inkling of Abbot's character showed up during, of all things, a board game. Dan Alderson brought the game to the Crest Building, a nearby expansion of JPL at the foot of the Angeles Crest Highway. Skip and perhaps Dave Hilt were the other players. The strategy of the game was that each player controlled a country. There were tanks and other weapons to wage war against countries. You could make treaties and agreements with countries, align yourself with someone against one country, anything goes.

All posturing and negotiating was verbal and non-binding. Dan made the statement that if you attacked him just once he would use his full power to destroy you. Abbot thought he could snip a few of Dan's tanks. When it was Dan's turn, he rolled the dice, over and over, as was your choice. Each roll of the dice destroyed one or the other's tanks. Gradually it became clear Dan was following through with his threat. Both Dan and Abbot were virtually annihilated. One surprising result of the game was that you could be severely reduced in your resources to wage war, but not easily eliminated from the game. It proved very difficult to totally destroy a country, so the game went on for a number of days. Lunch hour became very exciting. It was after all, just a game. But personalities were exposed at the same time. Abbot repeatedly broke truces. I was left with a lasting impression that Dr. Abbot could not be trusted. Al Joseph found the lunch hour game very disrupting and suggested it should end. The game was completed after work.

Mike one day mentioned that Abbot had received a JPL award for his work on the numerical integration package... whoops. How was that possible? First of all, Dr. Lawson was the father of the numerical integration package. It was considered my baby, so I suppose I was the mother. Who was Dr. Abbot? (A distant uncle perhaps) Of course you have to know that most of us took an in-house award like that casually. We all deserved it, but were not aware of it and probably could have cared less. Abbot must have researched the award and applied for it without telling anyone. Mike just laughed about it. I felt some immediate resentment, but it quickly passed. There was definitely something different about Abbot.

Dr. Abbot and I went to lunch together occasionally. A few times we went to downtown Los Angeles and passed out anti-war pamphlets. We shared the same strong feelings about the Vietnam War.

Victor Legerton was hired to help me write the documentation. This was almost unheard of among programmers. But this was an indication of how heavy the workload was. Victor was very friendly and helped relieve some of the pressure. He was a junior programmer and I always found technical writing very difficult. Mike usually handed back my first attempts at documentation full of red marks. I think he got tired of that and hired Victor. Just before I left JPL, Victor was laid off. He was unable to find work immediately. Since I was leaving JPL, I called him to let him know that there might be an opening at JPL. I was happy to hear that he got hired to work for Al Khatib on FASTRAJ. Mike had told me before I left that when Victor was laid off, Dr. Abbot in all seriousness had asked Mike to replace me with Victor. He told Mike that Victor was a much better programmer. This was ludicrous, and Mike and I had a good laugh.

The story does not end here--there is one last episode. I probably should have known better, but a couple of years after I left JPL I visited Abbot. I was hungry to find out how the Viking and Voyager projects were progressing. We spent a few hours recounting the remarkable performance of the DPODP. DPTRAJ was so reliable that it was referred to as the "gold bar" of the industry. Relativity was verified because the trajectories with the relativity model most closely resembled the actual trajectory.

Relativity verifications started in the early 1900s with the bending of starlight near the sun during a solar eclipse. Later atomic clocks in airplanes, subatomic particles and other exotic experiments proved Einstein correct. The DPODP represented the 7th verification according to Abbot and was unique in that no extreme circumstances were necessary. Ten thousand miles per hour or more is not exactly ordinary, but compared to the speed of light it is nothing

Another thing Abbot told me that he believed to be true was that the DPODP was the most accurate measurement of the speed of light, astronomical unit and the frequency of cesium. It is possible this was true for a brief period. Lasers from mountaintops are now used for the measurement of the speed of light(now a constant). We were prepared with relevant partial derivatives but experiments outside the DPODP would prove more accurate.

Abbot was about to make a pleasant evening's visit very strange. He remarked that Khatib's solar pressure model was not being used. He was sure I would delight in hearing that fact. But of course, nothing could be further from the truth. As I began to leave Abbot's apartment he got "syrupy" about my accomplishments. Somewhat like "you are the greatest." It embarrassed me. In retrospect, Abbot had always resented that he could not do the work I had done on the program. I think it was eating him up. I had no idea what was coming next. As we discussed plans to meet again, I climbed into my car. Motor running and the window rolled down, I said goodbye. Dr. Abbot said, "You know, John, I never really liked you."

I drove off without a reply. I hope Abbot is OK.

CHAPTER 26

BEER AND MORE BEER

That Mike Warner was a connoisseur of beer was an understatement. Mike's episodes with beer could probably fill many pages. I guess the best place to start is to tell you about Mike's wedding reception that I attended. There were kegs of beer everywhere. I don't recall champagne, wine, or anything else, but there was definitely plenty of beer. Mike's idea of the perfect vacation was to fly to Olympia, Washington and take the "beer boat" to San Pedro, California. The "beer boat" could accommodate approximately ten passengers. After all, the rest of the space was for beer. As a passenger he could drink all of the beer that he wanted. Mike allowed himself ten beers per day so that mapped out to fifty beers for the five-day trip. He always held himself to those restrictions.

At his home in Wrightwood, Mike decided he was going to plant an apple orchard. He felt the temperatures there in the mountains were ideal for producing apples. Apple trees need a very serious period of dormancy in the winter. He carefully chose the variety of apple. He had to dig large holes for each tree and put fertilizer and one rock under the roots of the new apple trees. His system was this: he laid out a row of full Foster Lager beer cans. As a tree was planted in the ground, he would then drink that beer, proceeding on to the next tree until the row was completely planted.

The drive from JPL to Wrightwood took an hour without a single stop sign. On this mountainous drive his Volkswagen would wear out a set of tires in eight thousand miles. Ordinarily they should have lasted at least forty thousand miles. Mike had decided it was because of the pea-gravel that was steamrolled into the blacktop. The pea-gravel gave the roads a bit more grip when there were icy conditions. One time I spun my VW on this ice. I never again went over thirty miles an hour at certain places on that mountain road. During the winter you never knew when ice had formed over night.

Mike's drive home from work included a six-pack of beer. He drank each can and finished it at just exactly the same spot each day. He had it all mapped out so that his last beer was gone just as he arrived at the outskirts of Wrightwood. There was usually no traffic on this drive. In those days the road was just ours. We could drive the whole way to work down the center of the mountain road. I had actually driven it many times and never met or passed another car. Occasionally Mike and I would carpool and I would drink a couple of his six beers. This was probably the only time I remember Mike becoming agitated. I had completely thrown off his schedule. When I finally realized this I started taking along my own beer. These trips were surprisingly productive. Mike was quite busy during the day. This was a perfect opportunity for me to run past Mike the daily events concerning DPODP. A lot of work was accomplished on these trips.

There were a number of contests involving beer and Mike. One of the most notable was when Skip Newhall and Dave Hilt put up a case of beer. Mike had to run a mile in eight minutes. Everyone knew Mike would win. We thought Mike could have even beaten Roger Bannister if enough beer was put up as the prize. However, Mike claimed he would never do that again. He collapsed at the end of the race and they were sure he was going to die.

Probably the most extreme beer episode was the night that Mike decided to drive Route 66 from Riverside to Pasadena. His goal was to have one beer at every bar he passed. He believes he had somewhere in the neighborhood of fifty beers that evening. He got into at least two fights along the way.

Mike was as brilliant a person as I have ever met, a veritable "walking encyclopedia." He was well versed on subjects that were really surprising. We discussed the upcoming Death Valley trip. He had charted the conditions and temperatures of Death Valley for us. He knew exactly the highest temperature ever recorded there. He knew all types of technical data about Death Valley. This was typical of Mike. He was knowledgeable about many different subjects. Mike could also write technically just like "ringing a bell."

CHAPTER 27

BAMBI VS. GODZILLA

Mike was over his head in personnel paper work, so he promoted a fellow that I will call John Robbins to manage three or four regular JPL employees. I was shuffled over to Robbins because I was a programmer in section 315. This was a peculiar arrangement, since I was a contract employee for Informatics but treated as a regular JPL employee.

John was well organized and a bit bureaucratic. I think I wasn't the square peg for his square hole and we did clash. He really couldn't be my personnel manager as I actually worked for Informatics. When it came to technical issues between John and me it was "Bambi vs. Godzilla."

We did not work well together. Somehow my enthusiasm for the work rubbed him the wrong way. John was a very sincere person. I felt we could have become friends, except there was always a certain distance between us. It was probably just a basic personality problem: two well-meaning people with the wrong chemistry forced to work together.

I was taking on more and more technical responsibility and John was put in charge of REGRES. REGRES computed the partial derivatives of the observables and clearly lagged behind all the Links on Mike's progress chart. I had analyzed REGRES, as I had all the other Links, and found it to be relatively straightforward. I did not understand why it wasn't ready to execute. It was essentially the light time solution and observables that were calculated. Give me a month and I would have finished it. This was probably why I sensed John's resentment toward me. I tried to be tactful, but there are times feelings are hurt no matter how careful you are.

I was working on a wobbling recursive mess and managing to keep my head above water. John's Link, REGRES, was on a stable well-defined platform and he was going under. From my perspective there was no reason REGRES should not be ready. I felt the whole program was being held up. Actually we would have far more work ahead of us than anyone imagined. REGRES wasn't holding anyone up.

I felt driven to see headway on the entire program, but was helpless to influence REGRES, and I guess I pushed too hard. I could sense a crisis was building between us. I believe that John never completed REGRES, as I don't recall hearing about any results until Dave Hilt took over.

At one point John did manage to get us four hours of daily-devoted 7094 computer time. This was a tribute to John's management skills and I was excited to get run after run each day. The regular daily operation gave you three runs in the best of circumstances, as you competed with everyone at JPL for the computer. Several times during stormy weather lighting strikes would damage the computer and set everyone back three or four days.

On the first day of devoted computer time, we all showed up with our decks, planning to use the four hours of prime time we were allotted. We waited for the operators to finish off the last of the batch programs and then it was our turn. Since I was the contractor, John decided I should be last. Each JPL programmer had his deck loaded, executed and printed in about fifteen minutes. I was next and my program loaded. To everyone's surprise one half-hour later it was still running. John looked embarrassed, so he decided to have the operators terminate my run. No one else saw anything but diagnostics. John was irked because my coding was the only program ready for this much computer time. I was pleased, but he couldn't run anything because his Link was still stalled in the coding stage.

Frank Newell laughed and told me that when I stepped out of the room, John said we could use my work as fill while the serious stuff was sandwiched in between. John either didn't understand how important my work was, or he didn't want me around. In fairness to him, John came to work on time. His demeanor was very professional. For the most part he let the programmers do their jobs. He made few inquiries about our progress. It was at this time that Mike Warner gave me the majority of DPTRAJ. PDQ was in DPTRAJ and represented a verification of Link VARY which was not part of DPTRAJ. Both programs calculated the same partial derivatives but with different formulations. PDQ could have been a very complicated Link. I evaluated it and discovered a new strategy that saved three man-years of work.

My success with PDQ seemed to be driving a larger wedge between John and myself. Regardless of what I did, I seemed to aggravate him, so I tried to stay out of his way. Outside of John everyone helped each other for the general good of the whole project. John couldn't admit he needed assistance. I regularly spent most of a day helping someone or vice versa. I believe his perceived position as manager prevented him from seeking help to elevate his level of competence. I lacked the relational skills to break into his shell. Interestingly, a book called The Peter Principle had just shown up on the shelves. Joe Witt read it and told me it was right on!

One day John began to take notice of my work. It was actually more than that--he began making suggestions about some of my methods. I disagreed and then he became more demanding. He was attempting to direct my technical work. Why he decided to get involved at that time is a mystery, as he clearly did not understand. He insisted I change directions in a way that clearly was incorrect. Try as I might he stubbornly refused to back off. I finally straightened him out. I was hopelessly committed to finishing the work. There was no looking back and I had become a gorilla. Nobody was going to mess up my program... nobody!

John kept a daily diary on his desk calendar. Dave Rose told me many years later he had noticed one day's entry that read: "Today he was openly contemptuous."

John Robbins went on to contribute to the language FORTRAN. He developed a structured programming pre-processor. This may have been the first example of top-down FORTRAN processing in the country. My understanding of his contribution to the FORTRAN language was to take the "continue statements" out of FORTRAN and eliminate some of the "go to's." He had asked me whether this would be any help in programming. I looked it over and concluded... not much. I must have been wrong, because his work turned out to be very important. These principles are now common in modern languages.

Mostly I remember that John wanted to be a chiropractor.

CHAPTER 28

DUMB DOWN

I learned a lesson about not demanding too much money for your services, even if you're worth it. I applied for a job near San Luis Obispo. It was with a small contracting company at a large aerospace and defense facility. When I arrived at the site it turned out to be a large white geodesic dome. I first had difficulty finding a door. It all looked very futuristic. Once inside I found ordinary offices and computers. The unusual part was the large and powerful equipment used to communicate with spy satellites. I was told it was dangerous to be outside when the satellite came by. The satellite gathered information over Russia then dumped it out over this dome. I don't remember if the danger came from the satellite transmissions or the dome transmissions.

The interview was with the main technical engineer. They had a problem that needed urgent attention. The satellite had been fed the wrong information. The main concern was about whether there was enough fuel to save it from crashing. I was immediately excited about the challenge. We started talking about my experience. I did look at some documentation and asked a few questions. When he replied, "state of the art" I returned "you must be using my program that I wrote years before at JPL." The way he answered led me to believe this was the case.

I was truly his best hope if anything could be done. I guess dollar signs must have shown up in my eyes. I have found that once you ink your salary on a contract, you will never be able to see substantial raises. Competition with fellow workers usually keeps you in a narrow salary range. Heeding the advice of the company manager about salary, we agreed on a figure that was above the level of his engineering staff. He was delighted with me and very excited that I might be able to solve their problem.

I went to Los Angeles to help my father while I fully expected to hear in a day or two that I should come back to San Luis Obispo and begin work. I did hear from them, but apparently a vice president back east at corporate headquarters had nixed the plan. She felt the higher salary was a problem for the regular employees. I thought about it and agreed, so I asked what salary they were suggesting. A day later I heard from a very disappointed general manager. Since they could not meet my salary demands she felt that I would jump ship at the next offer. Nothing could have been further from the truth, but the negotiating was finished and next week I went to work for another company, Reichert-McBain.

A few months later I drove up that way, hoping to stop in and talk with the technical engineer I first interviewed with. I was interested in what happened with the satellite. The offices under the dome were vacant. I guess that answered my question.

Then there was the politics at North American that surprised me. During a critical three-day period, I was asked to help a programmer who was going to be out of town. An engineer was scheduled for a presentation in Washington D.C. in two days. His specialty was error ellipsoids around in-coming ballistic missiles. Ellipsoids are sometimes shaped like footballs.

The missiles are most likely to be near the center of the football. The three-sigma standard deviation guarantees the missile should be somewhere inside the football 99.7 percent of the time.

I was instructed to run a program that calculated the major and minor axis of the ellipsoid along the missile trajectory. There were two types of missiles, and it was not hard to figure that these could have been Soviet.

The engineer needed these runs for his presentation at the Pentagon. There was a small hitch. Before the programmer left he said he found a bug in the program. Would I please fix it before submitting the runs? Looking over the coding, I found that it clearly had a blatant mistake. It was also obvious what the correction was. I ran the program with and without the bug. The error ellipsoids were dramatically different. The engineer was very puzzled. The runs with the bug gave him what he considered reasonable results, which is what he had been accustomed to. But, the corrected program gave results that he had difficulty explaining. He was an expert in this field, but what was he to do with his presentation a couple of days hence? I could not find any more obvious problems, although with a strange program and only one day to work with it, you wouldn't expect to turn up much.

I was very curious to find out what he would do. He sounded somewhat apologetic, but his plan was to use the old data with the bug. It supported his work and these were conclusions gained from many months of work. I never saw him again or ever heard what, if anything, was resolved. I still wonder what the correct answers were.

CHAPTER 29

PREDICTS

PREDICTS is a program which supports the DSN (Deep Space Net) tracking stations, but is not a part of the DPODP. The program supplies information about where to point the antennas so as to acquire the telemetry signals of the probe. Following the same pattern as the sun and moon, a distant space vehicle's electromagnetic signal will rise and set each day. As the Earth turns and the probe comes into "view" at the horizon, the tracking antenna must know where to point to look for the probe. PREDICTS supplies this "pointing" information by reading the probe ephemeris calculated by DPTRAJ. Using the probe's state vector, observable calculations similar to those in the DPODP give accurate predictions of the direction for the incoming signal. Once the antenna acquires the signal, positioning systems within the tracking station lock on and continue gathering data until the probe "sets" and the signal is lost.

With strategic tracking stations located around the world, a probe signal may be received in at least one location twenty-four hours a day. Microwave linkages direct the data to JPL in Pasadena. Rubidium crystals at the station time tag the observations, usually Doppler, for precise record keeping. But it is the large main frame computers in Pasadena running the DPODP that extract the information for navigation.

To obtain the maximum window of opportunity for receiving data, each station maintains horizon terrain silhouettes. These may be mountains or other signal obstructions as well as ionospheric effects. These considerations will not impact the accuracy of the data and are only used to squeeze out a few more observable measurements. Equipment malfunctions and multiple missions may also affect the reception window.

Similarities of the observable calculations within PREDICTS and DPODP made it a natural for inclusion within the DPODP. The rationale to exclude it represents an interesting aspect of the twists and turns of software development at JPL. PREDICTS is not a source of data for the DPODP and was therefore self-contained, requiring only the probe ephemeris as input. Splitting it off was most likely a strategy to help reduce the burgeoning size of the DPODP. I was not privy to the internal struggles which were to haunt PREDICTS. Success and failure of projects could vary broadly even within close knit groups. Turn a corner in the hall and you would find widely inconsistent software quality. I did wonder how virtually a carbon copy of segments within the DPODP could evolve in such different ways within PREDICTS.

The upper management at JPL was unquestionably the finest I have ever worked with. Their presence was only evident when we had a special need. No nut and bolt oversight--they let us do our jobs and trusted us. I have always wondered if the team of astrophysicists assembled at JPL in the late 1960s was unique and irreplaceable. Could another autonomous group of NASA employees have developed a successful computer system for Viking and Voyager? Decades later with a great deal more software experience I look back and am still struck by the incredible synergy of Sections 392 and 315.

Outside the umbrella of the DPODP programming group there were rumblings that PREDICTS was troubled. The criticism sounded much like that which you might have heard years earlier about our program. A possible explanation was that antenna and observable specifications change frequently and would make software development difficult. It may have been put on the back burner with other mission critical problems demanding more attention. Many years later (1985) I went back and visited JPL. Among other things I asked how PREDICTS had faired. To my astonishment I learned it was still a very tortured program. It was now coded in four different languages including COBOL (did someone from the business department come in to help?). There had been no progress in the past year; instead there were arguments about strategies on how to clean it up. Was the early DPODP susceptible to the same fate but somehow escaped? In spite of the seeming disarray, PREDICTS did serve a critical purpose and should receive unequivocal acclaim.

In retrospect, there was no single dominant person on our project. We brought a diversity of skills that complemented each other. Besides, we were all good friends. This is not to say there wasn't competition between us; often we tried to out work each other. On the football or baseball field all camaraderie ended. Larry Ross played with the Cleveland Browns in the late 1950s. He would not play football with us but at the soft ball games he could smash it a mile. I remember catching a couple of his drives. The football games did not go well for the programming group. I was quarterback but no one besides Dale Bogs could catch a pass. Mike dropped one that would have been a touchdown, otherwise the analysis group scored on every drive.

I left JPL and moved to Northern California in 1970. Since then there have been puzzling results from Viking regarding the question of life on Mars. The pictures of the surface were spectacular. Voyager would establish a new level of excellence for scientific discovery. The close up pictures of the planets and moons now adorn every schoolhouse. Striving to exceed this record may have contributed to an over-optimistic design, and in this manner delayed the Galileo mission. With the possible exception of man landing on Mars, I don't believe the excitement of Voyager will be exceeded in our lifetime. With increasing accuracy demands because of ever more sophisticated missions, DPODP would expand. Now referred to as the ODP, the past three decades have produced a succession of modifications and enlargements to this program.

In the late 1970s some of the algorithms derived explicitly for the DPODP were generalized to conform to industry standards. Fred Krogh developed a numerical integration routine, probably incorporating methods used in DPTRAJ. This drop in, black-box-type package represented an outstanding structural modification for DPTRAJ. This work by Fred provided the world at large access to these leading edge mathematical methods. His name was attached to this package and I extend a hearty congratulation for furthering this important work. The world of numerical analysis and general mathematical library functions, available on computers, was profoundly influenced by developments within DPODP.

Programming standards were instituted that reduced the "disaster" programs so prevalent in the 1960s. I continually wonder how strict oversight would have influenced my work.

Probably never again will anyone have personal responsibility for such a large and important piece of software. I thank JPL for that exciting experience.

FORTRAN, oh my precious FORTRAN, was slowly relegated to the "untouchable" class of languages. Manufacturers stopped maintaining it, and it became the "Latin" of scientific computer languages. The ODP continued 100% FORTRAN to this day. Conversion of DPODP to the modern FORTRAN is underway. In addition there is a C++ "ground up" project underway, but preliminary reports indicate it will be a number of years before it is completed. What a daunting task. Now the three hundred thousand FORTRAN statements of my era have swelled to several million! Computer accessibility and reliability are significantly improved, from the early days.

Thirty or forty engineers are the current staff size. Astonishing accuracy has been achieved. By 1970 a station's location on the Earth could be determined to a few inches. Currently it is down to a few millimeters and dropping. This allows geosyncronous satellites and global positioning data to monitor continental drift and fault stress. Earthquake science is improved. Earth tides are now part of models, allowing measurements of the Earth's movement due to lunar cycles. Comet tails have been added to the drag model. It is exciting to think about the recent JPL probe that has managed to scoop up a bit of a comet and is returning it to earth.

CHAPTER 30

JACOBI SYSTEMS

Prior to my move to northern California, Alva Joseph, Ken Taylor and I started a computer corporation called Ultra. We maintained regular full time jobs at JPL, so in a sense we were moonlighting. We usually met at Al's apartment in Los Angeles; occasionally we got together at my house in Sierra Madre. This is where my neighbor, Bruce, heard of our new business and joined one of our meetings. He worked as an insurance agent at One Wilshire Blvd in L.A. (the Miracle Mile). His father, a part owner of the Sheldon Insurance Co., handled profit sharing programs for small businesses. This was the late 1960s and for profit dividend disbursement, computers had not as yet come on the scene. Every company had unique needs that made preparing the annual statements a long and tedious hand process.

Ultra was retained by the insurance agency and provided computer support for their profit sharing business. This gave Ultra an encouraging beginning and provided us with a small cash flow. I wrote the profit sharing program while Al and Ken did most of the work running it. They rented time sharing computer facilities and discovered that the nuances of profit sharing frequently required program modifications. Profit sharing for small corporations was often a uniquely planned scheme intended to impress the employees. This made it very difficult to design a general program for use by individual companies. In the end we found each company required a special program. The profitability of Ultra, which had excited us at first, began to dim. The time required tailoring each company's requirements to a computer meant slow growth and a smaller margin. On to the next challenge.

Our first big break came when Al located a very good contract software project in the San Fernando Valley. Al's friend had organized a similar small computer corporation and was already on board at a company called Jacobi Systems. This was a Navy project and the prime contractor was Varian Systems, a large eastern firm. The computer system we would work on was called VAST (Varian Avionics System Technology).

This was leading edge Defense Department technology, but most likely would also have significant civilian applications. The Navy wanted a system whereby electrical engineers would be able to automatically configure electronic circuitries. This would take the form of a two-step compiler allowing electrical engineers to specify component arrays that were configured as testing instruments. The variety of allowable components were to become the workhorses of modern electronics, i.e. resistors, transistors, diodes, capacitors etc. It was speculated that a system such as this would help speed up the turn around time of carrier aircraft during military operations.

This type of computer language had been the dream of Navy brass for many years. The inadequacy of second generation computers kept it on the drawing boards, but not out of the minds of the military planners. The new UNIVAC 1108 gave promise that it may now be feasible. A two step compiler was envisioned which would first syntax and semantic check the engineering coding on the 1108 and pack it into files for acceptance by the smaller onboard real time computer. Our entry in the process was seen by some at Jacobi as critical in determining success or failure. Everyone was by then aware Jacobi was behind schedule and in trouble. There was as much activity in the evenings as the mornings, because late night contractors were swelling the ranks. Although there was no panic, I sensed that significant progress needed to be demonstrated in order to keep the prime contractor, Varian, from moving the project back to Langley.

It was an interesting problem, especially since I had never seen the inner complexities of a compiler. The UNIVAC FORTRAN compiler was increased from second generation compilers to allow ten thousand symbolic names, which was the single most significant upgrade that made possible this new electronic technology. Our front-end compiler would be programmed in FORTRAN, and the new language for the electrical circuits would reside in the 1108 as FORTRAN symbolic names. This cleverly connected our coding and addressable storage. The FORTRAN compiler internals provided some assistance validating our new language but a much deeper syntax and semantic check would still be necessary. This was our job for the next six months, and it entailed forty hours of work per week on top of our daily JPL responsibilities.

The smaller on-board computer did not have the capacity of our large main frame, so we had to be sure there were no language violations when we packed the character string of instructions. I could characterize syntax and semantic checking as some of the most delicate work I have ever seen (syntax akin to spelling and semantics akin to grammar). Besides being the sheriff of the language, our diagnostic messages must also advise the electrical engineer about alternatives if he violates a rule. These run the gamut from terminating his run to warning about possible consequences and continuing the compilation.

Internal consistency of the language is another complex issue. We asked ourselves, is it possible to adhere to the language rules and still be sustaining contradictory results? I discovered a difficult circumstance with a floating-point character string. From the moment the programmer entered a number on the keyboard to the final packed records sent to the shipboard computer, several discrepancies were uncovered. A floating-point number may be an addressable word or a character string in memory. Converting back and forth between these data types could cause the exact value of the original number to be lost. Rounding and significant figure system issues make determining the original input uncertain. The binary representation of 1.0 and 0.9999 illustrated one of these issues. Electrical tolerances (ohms, amps, watts, etc.) for the components and accurate representation of the input in diagnostic messages could possibly be compromised by losing the exact character string keyed in by the programmer. These seemed relatively small annoyances to me, but were in fact important for the integrity of the compiler.

This revelation gave Jacobi a chance to demonstrate to Varian that we were on our toes since they had overlooked this contradiction in the original design. My colleagues at Ultra were also making important contributions, and Jacobi was pleased with our work. Ken helped me interface with difficult parts of the system, and gradually progress was emerging. We could now begin to test our operational performance. It was all very exciting.

As the months wore on we slowly adjusted to working eighty hours per week. By nine in the evening we often went to a fancy restaurant and finished up our work at around midnight. I then drove 100-mph on the L.A. freeways to get home and tried to get as much sleep as I could before 8 am when I went back to JPL.

The news arrived that the Navy brass was scheduled for a visit. I sensed we were probably prepared. I was delighted when my coding was selected for the demonstration and was sure it would go well. I caught the flu and was going to stay home. The management at Jacobi said I had no choice, I had to be on the premises when the Navy was there. So I spent the day lying in my car with 104-degree temperature. A couple of Navy officers came by to commiserate with me.

The demonstration was flawless and we were asked if we wanted to move back to Langley, Virginia for follow up work. I declined, as I was still heavily involved in the navigation system at JPL. The salary would have been double my JPL pay, so it did tempt me. A few years later I saw a General Electric advertisement in Scientific American touting a system called VAST. There it was, a picture of America's most advanced aircraft carrier. I felt satisfied that our hard work was successful.

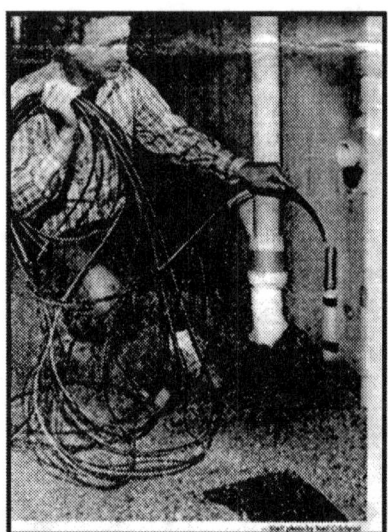

Above: (Monday, August 3, 1981)

Below: (Monday, October 23, 2000)

Courtesy: Times-Standard, Eureka CA.

CHAPTER 31

CLAM

(Computers - Lasers - Atomic clocks - Microprocessors)

Although a long way from my previous work at JPL, the subject of earthquakes would become my new focus.

There are a variety of physical changes purported to be associated with earthquakes: animal behavior, subterranean gas, electromagnetic waves, planetary alignments, plate tectonics, movement of faults, lack of movement of faults, and a variety of pseudo-scientific effects including headaches in human beings. Some of these have been established as probable precursors of earthquakes. There has not emerged existent prima-facie evidence of an impending earthquake. Statistically we know that major earthquakes tend to occur in the same general vicinity, possibly hundreds of years apart, and that death and destruction are a guaranteed aftermath. Large earthquakes usually produce smaller aftershocks with additional damage. Small microseism quakes sometimes precede larger shocks. Volcanic seismic activity is generally regarded as a separate science compared to traditional earthquake fault rupture, having its primary cause upwelling magma.

The Flux Capacitor

Where does an idea come from? Is it a vapor, or a bolt out of the blue? It is hard to remember exactly when I came up with this particular idea. Possibly it started with a visit from Mike Warner. Having given up on the fast paced lifestyle of Los Angeles, I escaped to the mountains of northern California. Mike managed to locate me even though I was twelve miles from the closest power and telephone lines. I am sure he was curious about my life in the wilderness.

Mike caught me up on the latest things going on at JPL. My last question was: "What would be the next interesting project in the field of science to work on?" "Earthquakes," was his answer. My curiosity was aroused. Thus began a deep continuing thought process about earthquakes. What could I do to improve man's understanding of them?

Leaving the wilderness and moving north to Arcata, I became a physics professor at Humboldt State University. Mike arranged an invitation for me to the Earthquake Affiliate Conference, the most prestigious conference at Cal Tech. The hosts, Drs. Charles Richter, Clarence Allen, George Hausner and Jennings, led vigorous discussions on the status of earthquake science. I had the opportunity to meet with them after the conference.

Back at HSU I was excited about this new field and began my study of the dynamics of earthquake science. I couldn't shake a comment I heard at the conference: "Next year we will only have more motion records. We need improved theories to explain the current data". I began wondering about other methods of measuring earthquake response.

Gradually the pieces of my musings fit together. I had it! Computers, Lasers, Atomic clocks, and Microprocessors (CLAM). My thoughts had gone from motion data that quantitatively surpasses all other recorded phenomena connected to earthquakes, to wondering if there could be some other manifold measurement. Trying to imagine how the motion was translated from hidden recesses of the earth to a seismograph, it occurred to me that motion was connected to pressure. Is it possible more information might be available in a pressure recording?

If any progress was to occur with prediction it was in understanding the triggering mechanism deep within the Earth. Measuring pressure in rock was my next mountain to climb. It was clear to me this should take advantage of the most important modern high tech developments available. Computers would be essential, but the key was the interface between the rock and the pressure-sensing instrument.

Since it was sensitive to pressure, I chose fiber optic cable as the sensor medium. The index of refraction of the glass responded to pressure and was inherently part of the technology of fiber optics. The lasers produced coherent light, the microprocessors and atomic clocks could theoretically measure the speed of light. (The atomic clock was one way to measure the speed of light, later I learned an alternate "interference" method might be just as good and save the expense of a clock). The earthquake could change the index of refraction of the glass by virtue of its pressure wave. This change could be studied by recording the value of the speed of light. Here was a truly novel high-tech approach to sensing earthquakes. My first impression of the idea was that of a very modern solution to a very old problem and I was excited.

The next consideration was how to get the glass fiber in contact with the rock or ground. This junction was a key to translating the pressure in the rock to the glass fiber. University expansion plans for a new science building offered an opportunity for a prototype instrument to be installed in the cement foundation. The building could enclose the entire experiment and confine the equipment in a secure laboratory environment. Everything was coming together.

Dr. Donald Lawson, head of Institutional Research at HSU, generously offered his assistance. Don and I would collaborate closely on all details of the work. He provided me with an office, telephone, secretarial services, and funding. His position gave credibility to my work and he provided very sincere support.

My list of contacts were a "Who's Who" in earthquake science: Drs. Allen, Richter, Hausner, Jennings, Urhammer and Bolt. In addition my former associates, Mike Warner and Al Khatib from JPL, were interested. Dr. Chris Harden from Cal Tech offered guidance. He designed the electronics and also estimated the cost of the project.

I purchased the fiber optic cable from Belden in Illinois, who also recommended other experts in related fiber optic technologies. With each additional contact I was usually supplied more leads and this led to a larger base of information. The State Seismic Safety Committee was interested and donated standard motion equipment for the HSU Science Building. This would augment the fiber system and provide comparison motion data.

I was ready to proceed. The responses to my newsletters were encouraging. There was no indication of any similar ongoing research and not a single negative reply to any of the inquiries I had circulated. Dr. Bruce Bolt, a top earthquake expert in the world, wrote me a letter of encouragement. Dr. Urhammer felt the issues I was raising were valid and should be investigated.

The practical issue of placing the cable in the cement foundation in the new science building was next. Dr. Lawson gave me a set of blueprints. We planned on installing the fiber bundles in two seventy-foot sections at right angles. This was compatible with the general "L" shape of the building. I felt it might be of some advantage in determining the direction of the incoming wave.

I stretched the cable out along the hall of the Physics and Chemistry building, measured and cut it. Next I proceeded to lay the cable on top of the gravel where the foundation would be poured. The workmen were alerted to do their best to protect it. One evening a friend and I entered the finished construction site with a laser. In the dark he sent laser light into the rough bundle of fibers. The twinkling I saw at the opposite end confirmed that the entire fiber was intact. These tiny hair-like sections of glass will remain within the HSU Science Building for its lifetime.

University Politics

I was about to learn a hard lesson. I felt I had been careful to get full approval for the project. The president of the University and the Dean of science supported it. The head of institutional research, Dr. Lawson, provided me with his facilities to use as needed. Several colleagues in the physics and Computer Science departments had contributed supporting letters, and Dr. Lester Clendenning collaborated with me over some of the theoretical principles. I did notice, however, a lack of interest from the Engineering and Geology departments. I spoke with a couple of engineers who taught courses in earthquakes and did not feel they quite understood the principles I was investigating. They wrote no letters pro or con, nor did they express any opinions at that time.

What went wrong? It was not my assiduous attention to detail. I had many leading world experts contributing to the research. The institution that stood to gain the most was going to eat its own tail. Perplexing as it was there appeared to be a wide gulf between "men of science" and science instructors. The first untoward contact I had with the HSU Engineering department was a request to present my ideas to a group of faculty and students. This seemed harmless enough, and since I had hoped for a broader interest in the project I agreed to an hour of discussion and questions.

The meeting was awkward. The faculty was not in tune with the physics and it was way over the heads of students. I presented the principles fairly well, but hardly any relevant questions were asked. There was a question as to whether it would "work," and my answer was "we won't know until it's built". There were no follow up questions, so I assumed they were satisfied. I did sense an imperious attitude by someone from the Engineering department. There was a gathering storm and I was going to be cast adrift.

Funding sources were suddenly the number one priority. This didn't make sense, since the state seismic safety group would install standard motion equipment and Belden volunteered more free cable. I had contacted every sector of the administration that might be able to help arrange funding. They were hard to catch in their offices. I persisted but in the end, after their many promises to pursue grant proposals, I believe not a single line was ever typed in any serious way.

A slowly developing turmoil was engulfing the project. The first indication of trouble was a short memo to Dr. Lawson's office from the Engineering department about concerns the CLAM instrument "would not work." This caught me off guard. Certainly we could not determine ahead of time its usefulness. I had always encouraged anyone interested in the project to contact me with questions or concerns.

The memo continued: one of the engineers had been poking around the foundation of the new science building. He found the protective metal conduit that surrounds the ends of the fiber optic cable as it emerges from the cement foundation. From this he concluded the metal conduit must transit the entire length of the building. (Actually the conduit is only anchored a couple of inches into the concrete and goes no further.) This erroneous conclusion led him to speculate that during an earthquake the fiber would flop around inside the conduit. I found his conclusion consistent with earlier conclusions I reached at the meeting, namely, that everything had gone over his head. My responding memo explained his misinterpretation and again requested we meet to discuss additional questions he might have. The "flopping inside the conduit" memo eventually disappeared from my file. Intrigue was now another detail to think about.

I tried to contact a newly hired HSU Seismology Ph.D. from Berkley. After numerous phone calls weren't returned and she failed to show up for appointments at her office, I concluded she didn't want to speak with me. The HSU Engineering department contacted a presumed expert in seismology at the University of Santa Cruz. He critiqued my last publication and in very strong words stated it could not possibly work. Did these people have crystal balls? The hot shot went on to say I didn't know what I was doing. The tone of this letter was clearly insulting and things appeared to be escalating into a personal affront. I remained steadfast, since building the instrument was the only way to test the hypothesis.

Don and Lester were still encouraging me, but the technical load was entirely on my shoulders. Don wrote a very strong letter to the engineers and clearly explained they did not understand the principle and purpose of my work. The stock clerk for the Physics department wrote a peevish letter reasoning the instrument couldn't work. Things seemed headed for a confrontation, yet no one contacted me personally. Why did this research threaten them?

I am sure we could have settled our differences, but they still made themselves unavailable. The world's experts were in my corner, correct? The president of the university was soon pulled into this ruckus.

A hearing on the merits of the CLAM system was scheduled. I had no choice but to publicly defend my work. The last straw was an unrequited demand for my attendance at this hearing. Six months ago I would have welcomed this opportunity, but now suspected an ulterior motive. On the defensive with the gritty truth emerging, an influential adversarial group was challenging me. Antipathy was going to be my reward.

The credibility of the university was now the issue, not my right to complete the research. It was actually suggested that I might embarrass the school if the darn thing didn't work. Don was now sandwiched in the middle and the battle line became the university president vs. yours truly. My first reaction was "they couldn't do this, can they?" Don assured me that I had no choice: submit to an academic roasting or the project would be terminated. Banished? I wondered how they would remove the cable from the cement.

Needless to say, this whole incident soured me on the world of higher education. I find it ironic that darkness and ignorance can be found in places that you anticipate light and intelligence. Provoking me into becoming a spectacle was hardly going to resolve any of the technical questions I had raised. Building the computer-synchronized speed of light contraption I imagined was the only way to decide these issues.

At that moment, I decided that the ensuing battle was not worth it. I still believed in the validity of my concept, but felt that it would have to be proven by someone else. Indeed, I was satisfied that I had canvassed enough of the world's experts on the subject. The word had been spread, and I was now ready to bow out. Others would have to carry the flag.

Vindication

Fortunately, some individuals did pick up the banner. Ten years later a patent by someone else in Los Alamos would confirm my hypothesis: the darn thing did work! Twenty years later Dr. Urhammer told me that Lawrence Livermore Laboratory was experimenting with a large spool of fiber immersed in silicone oil. Embedded in crushed rock, this serves as an earthquake sensor. Corning, in 1985, attempted to glue fiber and rock for a similar purpose.

All of this has little bearing on the quality of education at HSU. My daughter, Kim, received an excellent education in Microbiology, and she is now working with genetically engineered drugs. Many fine instructors are doing a credible job assisting young people along life's path. Some of my problems may have been the result of high stress levels, but in reflection, and with some sadness I have to say I found very few gentlemen at HSU.

My "Flux Capacitor" does remain imbedded in the HSU Science Building foundation, awaiting an inquisitive mind.

CHAPTER 32

IT WAS UP TO AZUL

Charles Gordon, a good friend of mine, lived in a small trailer above Redwood Park in Arcata, California. Charlie and his wife Sherry were librarians at HSU. He suggested I read Clayton Koppes' book on the Jet Propulsion Lab. JPL and the American Space Program was the first historical book written about JPL. Koppes' depiction of JPL ended at about the time I arrived at the Lab. Since I had worked at the Lab, Charlie knew it would interest me.

It is his pet parakeet, "Azul," that this story is actually about. Spanish for blue, Azul was naturally a very pretty blue bird. He sang loudly and sat on Charlie's shoulder while Charlie watched TV. One day, walking out of the trailer, he forgot Azul was on his shoulder. Off into the redwood forest Azul flew. The more Charlie coaxed, the further Azul flew. As darkness fell, Charlie returned without his friend.

The next day Charlie related to me the sad predicament. Three days later as I was walking around Arcata's main square, a small group of people were gathering around a red cedar tree. Sitting on a branch about fifteen feet away was a small blue parakeet in an obvious state of agitation.

My adrenaline went up a notch as I recognized my responsibility to rescue this bird. Assessing the situation, I sent my son Eric off on his skateboard to call Charlie. How was I to catch this bird before some harm came to him? I was sizing up the situation as Azul decided to change to a smaller tree, barely six feet tall. I sauntered towards the tree as Azul broke. He was fluttering not more than five feet above the ground as I sprang into action. With my hand outstretched I was gaining on him, when he abruptly changed directions. The bird went one way and I went another, as bystanders cheered him on. I discovered a bird has more turning power, i.e. wing vs. air, than a man has, i.e. tennis shoes vs. hardtop. I was again within two feet when Azul made a change in his flight plan. As Azul turned, I went straight ahead as the crowd continued to cheer. He returned to the original cedar tree, flew behind a burger business, and then into a parking lot. I resolved then and there to follow the parakeet... no exceptions, no conditions.

Nonchalantly, Azul flew over the fence behind the parking lot. "Let the adventure begin!" I yelled hastily to the crowd. Two young men grasped the emergency and took off with me. The next fifteen minutes was a frantic chase, probably similar to a Charlie Chaplin police movie. I felt the young men were not being aggressive enough when they had a shot at Azul. The bird was on the move and we climbed our first fence, crossing backyard after backyard. Azul had a close call where a few feathers were swiped away. He then kicked in the after burners and flew away.

I felt we had probably lost Azul, but went ahead and made a sweep of the general area. I was surprised to find him in an apple tree. The rough bark and high branches was enough to dissuade me from climbing the tree, so I decided to wait.

This gave me a chance to gather my thoughts. Was Charlie ever going to find us? I wondered if Eric had found him. I was prepared to wait indefinitely for Charlie. There wasn't much we could do; it was now up to Azul.

As I gathered my thoughts, I resolved not to run into the street chasing Azul. I would keep him in sight and hope that somehow Charlie finds us. Here was a bird that had survived three days in the redwoods, crossed a freeway, and was over two miles from his home. Azul seemed to be calming down, and just sat perched in the apple tree. By now, my helpers were getting bored, so I thanked them as they took off. I guess Azul finally recognized me, because all of a sudden he flew down and landed at my feet. I realized this was my chance and literally dove for him. About half way to the ground I had a fleeting thought: "don't hurt the bird!" Azul was flattened, but unhurt. I cupped him in both hands as I headed for the downtown square.

Do you suppose he was grateful for his rescue? Albeit, I had pushed his face into the ground. Showing no gratitude the bird tried to bite, peck and scratch his way out of my grip. "Owe!" "Oh" "owe" "oh" as I ran with him, but I had no intention of opening up my hands. The cook at the burger hut quickly punched holes in a paper bag and I thrust Azul into it. An astonished Charlie arrived, I went home for a much-needed shower, and Azul went home for a large lunch of birdseed.

    (At least I think it was Azul).

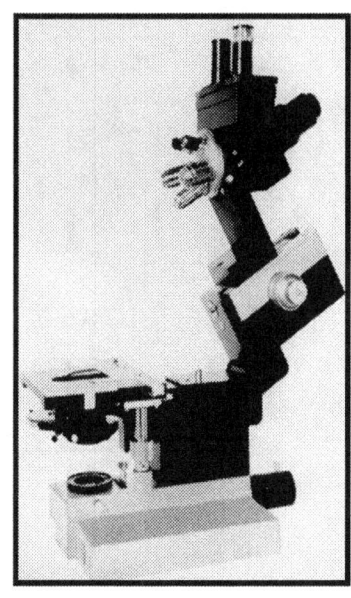
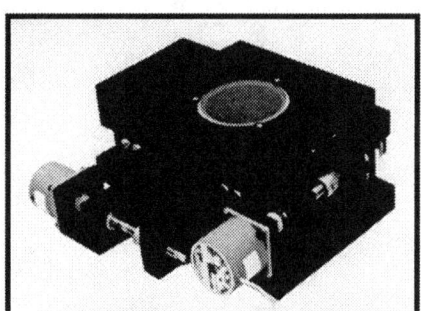
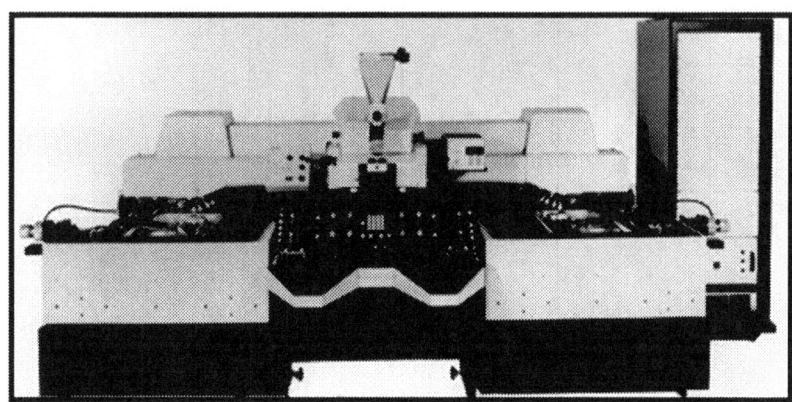

CHAPTER 33

CORNING AND REICHERT-MCBAIN

In 1985 I decided to leave academia, get a real job, make real money, and begin caring for my father in southern California, all at the same time. I landed in Chatsworth, the small high-tech business center in the San Fernando Valley. Among the one hundred employees five were recent computer science graduates. The balance of the company worked with hardware assembly or marketing.

Mr. McBain, a very competent electronics engineer, together with a very capable Austrian optic firm, founded Reichert-McBain. McBain would learn that technical prowess does not necessarily lead to management success. Originally, the company, which had a reputation for excellence, only manufactured microscopes. I believe McBain watched the ups and downs of his high-tech neighbors and wanted a piece of the action. Just down the street was the fifth fastest growing company in the world. Video cameras and computers were added to the microscopes, pushing the company into a new and very competitive technology.

Reichert-McBain would take a perilous leap into the world of computer chip inspection systems. The transition from microscopes to automatic wafer inspection systems did make some sense for Reichert-McBain. Wafers were at the heart of the manufacturing of computer electronics. A microscope was a key element in verifying the integrity of the transistors and circuitries of computer chips. This was quality control and to measure and validate the transistors on a wafer saves IBM the trouble of replacing faulty chips down the production line.

Evolving from a hardware background, Reichert-McBain reveled in the mechanical and optical quality of its new products. There was, however, lurking an unseen software crisis that would topple the company. What I learned about software after joining the company taught me humility, and led to the realization that I couldn't solve every computer problem.

Reichert-McBain had no previous software experience whatsoever, and believed they could just hire a few programmers. I watched one of our vice-presidents look at a floppy disk and wonder how it could cost so much. He would let the light shine on it. The iridescence must have been intriguing to someone who only understood the manufacture of hardware. He was a typical example of the upper management.

The accounting manager had recently graduated from business school, and set up the proper job descriptions for a company at least ten times larger than Reichert-McBain. Consequently, the business department was vastly overstaffed. Except for the personnel manager, the department worked an average of two hours per day. It was not uncommon to see a technician struggling to install a RAM board next to an office where the billing department read the newspaper all day. The sales staff busily wrote contracts without understanding the software development costs. Some of our customers must have realized that the contracts they negotiated were not realistic. But, they would hold us to the contracts, which were outrageous bargains for them.

The software functioned for routine situations. This pleased some customers, but often a new application created difficulties. Predictably, the customers requested changes. This played into the hands of outside contractors who controlled key source code. They were holding our company hostage and would modify the software (which was actually our property) only if they were given more money. It was up to us, "pay them now, or pay them later."

After months of inquiries I actually received documentation on our standard instrument software from the contractors. It was printed in pencil on both sides of a single three-by-five index card. This was delivered in a manila folder to make it appear more impressive. The cost to Reichert-McBain for the software represented by this card was several hundred thousand dollars.

Another group of contractors used a different ploy. They did relinquish the BASIC source code, but removed the coding comments. This made it nearly unintelligible, thus securing their meal ticket. The owner of this group dropped and broke one of our glass-scribed standards worth seventy-five thousand dollars, so he wasn't permitted in the vault. I questioned whether he should have been on the premises at all.

With all this going on, my task was to program a fiber optic quality control system for Dow Corning, the world's largest producer of fiber optic materials. It was a new system involving both software and hardware, which would speed up the monitoring of their fiber production. The software component of this system had been bid at twenty thousand dollars. My first impression was that they had lost a zero on the contract. The actual software cost to Reichert-McBain was well over two hundred thousand dollars and even that amount represented a miracle. The contract had already been inked when I was hired, so it had been suggested that maybe there was some give on the hardware pricing to make up the deficit. My attitude was "some companies give you the hardware, then charge you for the software. Software is where the profit is." Reichert-Mcbain had it all backwards, but I decided to leave the question of profitability to the business department.

Corning had a very interesting problem impacting a key element of the world's burgeoning telecommunication industry. The quality control procedure monitoring fiber production was not accurate enough for new products and in addition the current testing system required excessive manual steps. They wanted a fully automated system: place the fiber in a chuck, focus and measure critical physical properties of the glass all in less than two minutes.

The glass fiber looks like a strand of hair. It is normally coated with protecting materials that make it look like electrical wire. The glass is composed of two sections, core and cladding. Each section has a different index of refraction. The cladding surrounds the core or central portion. The core transmits laser light, which carries digital signals for telephones, computers, etc. The cladding surrounds and prevents most of the light from escaping the core. The difference in the refractive properties of the core and cladding is what confines the light to the core. Most light leaking from the core is bent back into the core, by the cladding. The bending is caused by the selective differences between the two indices of refraction.

To understand the quality control system we were developing, imagine a normal television monitor with a cross section of fiber magnified so that the cladding fills a large circle on the screen. The core is a small central circle about the size of a pencil eraser. A back lighting system creates a contrast between the circular core and cladding regions, which should have approximately the same center.

Several geometric measurements were going to be necessary with the new system. How close were the centers of the core and cladding? How near circular were these apparent circles? These measured tolerances would determine the quality of the manufactured fiber. Fiber optics had been invented several years earlier, and the Bureau of Standards was still working on developing universal calibration standards for the production of the fiber. Quality was clearly important for this emerging industry.

The basic testing equipment we were to develop used an optic system (microscope) to enlarge an image of a cross section of the hair-like glass fiber. Standard (non-laser) light inside and outside the fiber (backlighting) illuminated the structure of the core and cladding. A video camera converted the magnified image to a digitized computer record, which was available for numerical computation by a mini-computer.

The first issue we struggled with was inherent non-linearity of microscopic video systems. In other words, seven pixels may measure one micron in one region of the screen, and eight pixels represent one micron in another region. (I use the word screen to illustrate a principle because a monitor is not an essential part of the system).

By scaling pixels to linear measurements with a patterned glass standard it was felt we might be able with our video computer system to determine more accurate linear geometries of the core and cladding. There were off the shelf linearized video cameras but this could not correct the entire light path. The optics for magnification was also non-linear. The entire system needed linear precision, which was theoretically impossible with the combined individual hardware components.

The measuring process began with non-laser light illuminating the fiber. Magnifying optics pass the light to a video camera, which divides the image into tiny spots or pixels. An analog conversion determines each pixel's "shade of gray." The image is now a stream of numbers. A byte of addressable memory in a mini-computer stores the numeric for each pixel. The best quality components available would still produce unacceptable levels of non-linearity. To eliminate this problem we arranged to print photographically a tiny grid of metal boxes on a glass standard. To the unaided eye this was barely a speck, but magnified it was an orderly pattern of small metal blocks.

Pixels represent the smallest indivisible units from a video image that an electron gun sweeps into the memory of a computer. Mathematics can then be applied to these (pixel) bytes. The sharp edges of each small metal box produce a transition from dark to white light. An automatic focusing system attempts to produce the largest transition possible. The edges of a box are where the sequence of numeric pixels changes abruptly. The centers of the boxes are grid points whose locations will be used to indicate distortion or non-linearity of the entire system. All measurements of the fiber can be corrected or linearized by interpolating distortions from the nearest grid points.

For linearizing to work it was essential that all components that participate in measuring the fiber also participate in measuring the box locations. The location of the boxes was measured with the same total system that measures the fiber. This meant that the box locations were distorted the same way the fiber measurements were. The key to the linearization was that the box locations were known (we had them made), so any departure from these true locations was caused by the non-linearity of the system. It's a little like a watch that is exactly seven minutes fast. You can tell perfectly well what time it is if you subtract seven minutes.

At the manufacturing site each morning it was planned that a Corning technician would calibrate the system by mounting the glass standard where the fiber was to be mounted. Our computer system stored the distorted locations of our grid and then applied these corrections to the day's quality control tests. That, theoretically, was the way it was supposed to work. However, other technical problems were about to surface.

The mathematics to determine the core and cladding centers and their "out of round" was yet to be finalized. Corning had paid several "eminent" Swiss mathematicians to prepare a treatise on this subject. I found this paper to be flawed in seven different places.

Three points determine a unique circle, which is college algebra level math. What if you have one hundred points in a random halo roughly tracing out a circle? Where is the center? This is usually referred to as an over-determined system. Numerical curve fitting is necessary. What we theoretically wanted was a circle that minimized the aggregate distance to one hundred points. The treatise said "least square," but I noticed the math they presented was actually "least quartic." This was not a major point, but "where there is smoke there may be more smoke." I carefully dissected their analysis; everything was generally in algorithmic form until the critical last step. This was a statement that Fourier coefficients were to follow. These coefficients were the critical bottom line for a value judgment about the out of round of the fiber. These quantities were only referred to, not derived, and the conclusions remarkably incomplete.

Fourier coefficients may be very useful mathematical tools. They can emerge from various geometric figures and there were no references for them on the treatise. I began to research them and found several references relating to almost every simple geometric shape, but surprisingly not a circle.

I searched everywhere, including problems in textbooks, but found no circles. I was building a case for the inescapable conclusion that the treatise was a white wash. I prepared an alternate corrected analysis for each of the seven issues I raised.

The Corning team arrived from New York bright and perky, but a bit jet lagged. They had come to the West Coast to help speed up the work, and had no idea a buzz saw was waiting for them. I bought a new three-piece suit for our first meeting. The only issue we covered the entire day was my assault on their mathematics. Patiently I attempted to tactfully explain the situation. I realized mathematical participation was not part of our contractual agreement. We were only supposed to program their equations. However, I understood this was the centerpiece of a multi-million dollar upgrade at Corning. It was evident the success or failure of the entire project hung in the balance. I had no idea what repercussions might descend from corporate Corning if the system failed. Our profit, for the most part, depended on selling them the hardware. I knew they had paid for a piece of junk from those "eminent" Swiss mathematicians who, following this meeting, were never referred to again.

The Corning team was totally unprepared for my analysis and unfortunately the meeting started to turn sour. By noon they seemed to accept five of the seven mistakes. After lunch it was their turn to speak. I had been on my feet for four hours. The chief Corning engineer, who had seemed a bit sarcastic in the morning session, turned on me. My last two issues were not going to be accepted and our disagreements became clearly personal. I came close to losing my temper and my clean white shirt and new vest were drenched. I continued to insist that I was correct and stubbornly would not budge. Why would they want to cover up blatant mistakes? Perhaps it was just too embarrassing for them. The quality control system would never work without addressing these problems. They left the meeting in a huff. My boss told me to never ever do that again. What did he mean? I never asked.

There was lots of peripheral work for me to finish while Corning stewed. I tried to forget the botched analysis. Corning could complete the algorithm anyway they wanted. That became my attitude. Three months later Corning returned minus the sarcastic chief engineer. She had been relocated to a different project. The team now agreed with all seven of my points. (I am usually always correct when it comes to math.) The new chief engineer presented some analysis on how to interpolate linearization corrections, which seemed adequate. We all became friends and it was now my job to host them at lunch and dinner everyday for two months. Reichert-McBain was planning to make up part of the profit shortfalls in the contract by schmoozing them lavishly at expensive restaurants twice a day. Tough job, but somebody had to do it!

We never did find a closed form method to locate the center of a "least square" circle. The Fourier Coefficients were scrapped. I called Al Khatib at JPL and discussed it with him. He located some research on the subject at UCLA. It helped explain various conjectures about the problem but it did not help solve it. By this time I had enough of the uncertainties. I wrote a small program that crudely collapsed a grid of points near the center of the circle. This routine searched and found the center of the circles to any specified degree of precision. This was "hack and slash" at its finest. The number crunching took several minutes on the small PC and was impractical for the Corning manufacturing operation. But, this did validate the method we were going to use operationally. The morning I told Corning about my research their eyes glazed over and I don't remember them saying "thank you." Anyway, I had now verified that we were in the clear and they got a bug-free system. As the first hardware assembly was being shipped to Boca Raton, Florida, Corning hosted a large, friendly send-off party.

There were other strange things going on at Reichert-McBain that might help to explain their downward spiral. The software department had thin partitions separating us from the high activity connected to the hardware assembly. The walls and dropped ceiling did little to muffle the sounds of metal on metal. Drilling, pounding and other ancillary noise were usually hard to ignore. There was one sound I found excruciating. For some reason a large hardware assembly needed to be moved regularly. It weighed at least a thousand pounds sitting on metal skids. When it was pulled across the concrete floor the sound was like an amplified fingernail scratching on a blackboard. There were afternoons when they moved the assemblage intermittently for an hour or more. Some of our young programming staff trained for this by going to rock concerts on the weekend!

Another slightly disturbing phenomenon was the sagging ceiling. Once or twice a week ceiling tiles would unexpectedly fall. After a time I discussed this with the plant manager. He agreed something needed to be done. Ear plugs and hard hats... Good joke.

Finally, the ceiling work was put out for estimates. The day came when the low bidders showed up. We all moved our workstations elsewhere. The next Monday morning, bright and early, Chatsworth was warm, busy and the ceiling was trim. No more signs of the duct tape we had used to replace the tiles.

About three weeks later the ceiling began to sag again. Then the first tile dropped. Investigating, I peeked above the ceiling at the most amazing labyrinth of wire cables holding it up. Unbelievable! In order to level the main weight-bearing members, there was a random spider web of wires. These wires connected to major structures above the ceiling and were attached to any available mooring with a square knot. One can imagine what those low-bidders must have gone through that weekend. It would have been interesting to watch them work as each metal cable was tied and retied to get it just right. Electrical conduit, rafters, fire sprinklers, nothing escaped their relentless pursuit for a solid tie down. Perhaps they even used a level. As the building heated up and cooled down each day, the cable knots relaxed, gradually moving the ceiling back to its original sagging state. Eventually a ceiling tile dropped and we got out the duct tape. I wondered who these repairmen were, and what their next contract might have been. Perhaps, building dams, making tires, or possibly nuclear power?

Most of our company's entire hardware and software products were simulated on a three-foot by eight-foot cart. There were RAM boards which we called Focus boards, computers, optics, laser controlled positioning tables and Heidenhein glass scribed standards (sixty thousand dollars each), all conveniently portable. This equipment could be wheeled to various locations in the plant. One weekend the cart disappeared out the front door. Detectives interviewed everyone and could not determine who was responsible. Industrial spying and espionage had become a reality to Reichert-McBain. Thank goodness I had insisted they make copies of all our software for an archive. This was helpful in rebuilding our simulation system.

The struggle to produce quality software in this environment was monumental. I am not sure our president actually understood the difference between source and machine code. I tried to speak with him about it, but he replied "I don't want to talk about theoretical things right now." I had no choice but to toss out a PASCAL program that was the most wretched piece of software I had ever seen.

The "trickle" of poor management became a "gush." Reichert's former software manager wished to express condolences to me for accepting this job. He described the management as infantile and glad he was now working down the street. The plant manager announced that his neighbor said the programming language "C" was great; so from now on everything was to be coded in "C." Except of course, the work for Corning, who insisted it be done in

FORTRAN (This turned out to be the only software developed that year). Then he started instructing everyone to say nothing to the customers when it became clear that schedule deadlines could not be met. "Buy time...for what?" was my attitude. "The truth would still come out."

We were selling a few standard instruments, which helped the company's cash flow. But all the software disputes were causing the custom work to grind to a halt. The difficulties inherent in developing custom products were never appreciated. They can be very difficult to estimate because of the amount of work required, and therefore the costs can vary widely. I concluded that a recent high school graduate could have done a better job managing the company than the succession of company presidents that were to cycle through.

Reichert-McBain eventually hired a very experienced and capable programmer. I had worked with him on a Navy project many years earlier. "Too little, too late," but at least I now had someone to commiserate with about the software. He could see the hopeless situation we were in and left after a few months to open a Japanese restaurant.

With financial trouble on the horizon, Warner-Lambert purchased the company. Possibly it was because our products sounded as if they were related to medical applications. Warner Lambert did not have the technical base to understand our high tech wafer inspection systems. In spite of some vague parallels with medical technology, the real difficulties were that Warner-Lambert made mouthwash and we inspected computer chips. This led to an even less experienced and more dysfunctional management team.

Mr. McBain was terminated. Two company commanders showed up and both returned to the east coast with their "tail between their legs." Next they hired a Kodak executive to play president (Maybe someone who knows how to make cameras could help). His management plan was to sit in his office and talk to no one. I was the software manager and he refused to speak with me (Corporate transcendental meditation). Finally, he decided to pay exorbitant prices to bring in the outside software contractors and slowly axe the in-house programmers. In Warner Lambert's favor I have to mention that they were very generous with their severance money.

Oh, by the way, what eventually happened to the unscrupulous contractors? Well, there was a super secret custom contract for IBM. I believe it was ball bearings melting and forming junctions by surface tension with wafer layers. IBM wanted linearization of the video system similar to the Corning work. They also wanted a similar improvement to the table motion. I predicted we would need two separate systems, one for optics and one for table motion. The contractors insisted that one generalized system would work for everything. I was right (I was always right about computer matters). IBM had been given the usual hype. Delivery dates that could never be met. Capabilities that were non-existent. Anything remotely feasible was stretched to the limit. The decision to merge the optics and table motion into one general system lost three months work. The Kodak executive was canned and the goose that laid the golden egg for the contractors... it closed its door and the equipment was sold.

A fine microscope company that could not resist the allure of the high tech fast lane was out of business. Strand's rule for custom software: add a million dollars after you think you have already fattened up the contract.

EPILOGUE

What will the future produce? Could a more comprehensive historical record of space exploration influence future missions? I believe it will, and we will begin a new relationship with our planetary neighbors. Stars are, and will remain, an unobtainable destination of human exploration for eons. The ups and downs of robotic planetary missions will remain a source of excitement and disappointment for decades. Men and women will eventually surpass all mechanical exploration and settle the age old question of life on Mars. Our sister planet will then yield up her secrets. So much to do, so little time.

Before moving to northern California, I was offered a position with another JPL section. It was a surprise when an engineer came by and chatted with me one day. He previously held the position I was being offered. He warned me not to take the position because his previous boss had threatened his life. Passions connected to scientific work can be very strong and words are often difficult to retract. The human drama at JPL has rarely been written about. This information cannot be found in cardboard boxes, and fewer and fewer of us are left to tell the stories. So much to do, so little time.

As I began putting together my notes and preparing to write this book, I gave a few chapters to my pastor to read. Later he told me he was surprised I was still alive. He had been pastor of a church in Temple City that had members who worked themselves to death at JPL. I don't believe this human toll is well known. Doug Holdridge, who was my counterpart on the single precision program, was the first casualty I was aware of. After writing most of the SPODP his mind collapsed. I saw him wandering the gardens at JPL, a solitary figure; eventually he passed away in a mental institution.

JPL is steeped in history and a rich reward awaits future historians. This book will become only the third written so far with a historical vend. The JPL archival cellars contain many hundreds of cardboard boxes full of technical documents, now seemingly forgotten and yellowing. Where are the stories of the heroic flesh and blood people who gave us our national pre-eminence in space exploration? Looking back wouldn't it be interesting to know if Johannes Kepler shook the hand of Tycho Brahe and congratulated him for his planetary ephemerities? So much to do, so little time.

It is nearing forty years since I worked at JPL. In fact, daily we witness the success of the latest Mars and Saturn exploration. After President George W. Bush announced a future manned mission to Mars, I had a dream that I was there. Not with the President, not on Mars, but at JPL where the navigation system for this ambitious Mars adventure had been undertaken decades earlier.

There we were, this old group of scientists from section 392 and 315. As we sat reminiscing it was very clear to us all that our breakthrough in celestial mechanics would never be denoted by a name such as Kepler, Newton or Einstein. No, it had been a pure team effort. A perfect navigation system was not possible or for that matter ever could be possible. Our work in the 1960s was a perfect fit for the dreams of American technology as it existed.

With laughter we recalled the fun and pranks that went on, helping us to release some of the pressure we were all under at the time. We remembered the impassioned discussions that always led to a finer edge for the navigation system.

We did wonder if the new generation of astrophysicists realized what a solid foundation we provided for them to build upon? Our heroics, pushing the envelope of mathematics and computer technology, could never be fully appreciated. But this did not matter, our reward was in what it produced: the pictures and scientific information describing the solar system found in thousands of books, along with multitudes of media chronicles representing the finest recorded example of our resplendent information age.

What of faster better cheaper? Funding should never have been an issue for this important work. If the Federal government can't come up with enough dollars to sufficiently fund the American space program, the private sector should be mobilized. Think of the enormous benefit to everyone in the world, and at such a very small price. Ask Bill Gates to help; after all, Bill, where would you be without aerospace?

Men have landed on the moon.

Spacecraft have landed on the planets.

Three probes have left the solar system to fly - perhaps for eons - toward distant stars.

I helped to steer them on their way, to chart their course.

APPENDIX I

The Grand Tour Trajectory

by Alva Joseph

August 1, 2003

Al giving Bobby Fischer a game

In June of 1964 I left TRW Systems, in Redondo Beach, California to go to work for Jet Propulsion Laboratory (JPL) in Pasadena, California. This job change was made because I wanted to be more involved in the United States Space Program.

In 1961, Michael Minovitch of JPL speculated that a spacecraft could be launched from one planet toward another planet in such a way that the gravitational field of the second planet could be used to bend the path of the spacecraft in a way that would reduce the flight time to a third planet. A similar flyby would be done at the third planet to go to a fourth planet, and so on from planet to planet. This speculation by Minovitch eventually led to investigations at JPL for space flights to outer planets that resulted in two Voyager Spacecrafts which flew by the giant planet Jupiter on the way to Saturn, Uranus and Neptune.

This article is being written to describe the trajectory work that computer programmers did to make the Voyager Project a reality. In the mid-1960s, I don't believe that there was a computer program that could do an "n-planet" Multiple Planet Trajectory in a single computer run. I believe that I was the first person to create a computer program that could run multiple planet trajectories over a range of launch opportunities in a single computer run. This article will present information about the Grand Tour Trajectory computer program that was developed at JPL which was started in 1964 and was completed in 1966. Some of the other people who played important roles in making the JPL Grand Tour Trajectories a reality will be identified.

My first assignment at JPL in 1964 was with a small, computer programming group headed by Raul Y. Roth. The department manager for that group was H. Fred Lesh who hired me. An engineering group headed by Elliot ("Joe") Cutting and Francis Sturms had a desire to create a computer program that could do Grand Tour Trajectories. The engineers in Joe Cutting's group had been attempting to do grand tour trajectories by doing hand calculations. The grand tour trajectories were much too difficult to do by hand, in a reasonable amount of time. One of the engineers from Joe Cutting's group, named Ronald J. Richard came over to talk with Raul and me about the possibility of creating a computer program to do Grand Tour Trajectories. I told Raul Roth and Ron Richard that I thought it was possible to create a computer program to do Grand Tour Trajectories by using some numerical methods and recursive subroutine techniques that I had previously developed at TRW.

Following is a statement from the book called "Voyager Tales" by Dr. David W. Swift where Dr. Gary Flandro defines the quantity called C_3 that will be used frequently in this paper:

"In simple terms, C_3 is the kinetic energy of the spacecraft relative to the sun as it leaves the gravitational field of the Earth per unit spacecraft mass."

The reason I was confident that a Grand Tour Trajectory computer program could be developed was because I had developed and used the following three recursive subroutines, in ways that were applicable to this new application:

1. Parametric Study Subroutine

2. One Dimensional Search Subroutine

3. One Dimensional Minimization Subroutine

I knew that these subroutines would be helpful in developing a self organizing system for a Grand Tour Trajectory computer program. These three recursive subroutines were written in IBM assembly language for the IBM 7094 computer. Each subroutine used a table (or sequence of machine cells) where input parameters and the subroutine's related calculations were stored. The address (or location in the computer memory) of the subroutine that calculates the function or dependant variable (i.e. C_3) is one of the inputs in the table. The address (or location in the computer memory) of the independent variable or variable to be manipulated (i.e. T_F) was also stored in the table. Everything that the subroutine needed to perform properly was stored in the table. Since these subroutines are recursive, the address of the subroutine itself may be stored in the table; when this is done the address of a new table must be given for the recursive call, when the subroutine calls itself. A recursive subroutine is also reentrant. If recursive subroutine A calls subroutine B and subroutine B calls subroutine A which is already running, then subroutine A is reentrant.

The input for the Space Research Conic Program (SPARC) allows the user to specify a range of times of flight, to investigate at the starting or initial planet. An initial time and an end time could be input. A time increment could also be input so that SPARC would run trajectories for all times in the time range, by starting at the initial time and successively incrementing the time by a time increment until the end time is reached. The range of times is defined in equation (1) as follows:

$$T_{Fi} = T_{Fo} + \sum_{i=0}^{n} i \Delta T_F, \quad T_{Fo} \leq T_{Fi} \leq T_{Fn} \qquad (1)$$

where T_{Fo} – initial time

T_{Fn} – end time

ΔT_F – time increment

Note: The summation denoted by the symbol $\sum$ in equation (1)

above should have i=0 below it and n above it. This means that

the summation is done for the following values of i: i = 0, 1, 2, …,n .

A grand tour trajectory starts at planet one (P_1) and there are as many as four times of flight to get to planet two (P_2). For each time of flight from P_1 to P_2, there are as many as four times of flight from P_2 to P_3, etc. Figure 1 below shows part of such a tree structure.

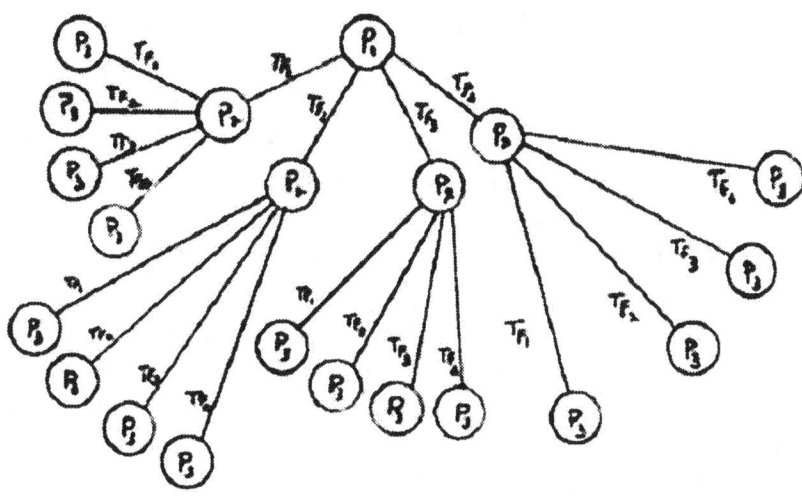

Figure 1

Figure 1, shows the "Tree Structure" where each branch is created for investigating Grand Tour Trajectory possibilities. The recursive and reentrant capabilities of the "Parameter Study Subroutine" made it possible for SPARC to be a self organizing system, since the various times of flight are not known before they are determined by the computer program while it is running. As the times of flight are found at each planet, the program puts them into a unique table for the "Parameter Study Subroutine" to fly the trajectories for each of the times of flight, from one planet to the next planet. This means that the Parameter Study Subroutine supplies up to four independent variables (i.e. T_F) and runs a patched conic for each, in order to obtain corresponding values of the dependent variable (i.e. C_3).

Ron Richard wrote the "Request for Programming" (RFP) to create the Grand Tour Trajectory Program. The name that Ron Richard selected for the computer program was "Space Research Conic Program" (SPARC). The RFP contained the engineering technical information that was needed for developing the computer program. The program structure and the numerical methods were left for me to develop. Fortunately, they allowed me to develop the program without having to write an extensive detailed design document for everything I planned to do; it would have taken a long time for me to explain everything in detail. I had previously developed programs at TRW that were similar to what was needed for the Grand Tour Trajectory program. I was an experienced programmer and I knew how to use the programming language as a design tool; this is usually a hard sell to managers. There are times when it is easier to do a programming task, than to explain what you plan to do.

SPARC would have the capability of flying a spacecraft past planets so that the gravitational field of the planet could be exploited to increase the velocity of the spacecraft, and alter the path of the spacecraft in a favorable way to get to the next planet, thereby reducing both launch power requirements and flight time. We assumed that the spacecraft would leave a planet with the same energy (C_3) in which it arrived at the planet. This was done to preserve the conservation of energy relative to the planet. To make the program run efficiently the RFP requested that conic sections be fitted between one planet and the next planet to simulate the trajectory path between two planets. An alternative would be to use the JPL "Double Precision Trajectory Program" (DPTRAJ) to numerically integrate trajectories between two planets. Obviously, the numerically integrated trajectories would be orders of magnitude more time consuming than fitting a conic section between two planets. The RFP requested that the JPL Ephemeris Tapes be used to get positions and velocities of planets, at desired epochs (or times), as needed by SPARC. The RFP also requested that SPARC would print the "Orbital Elements" and "Energy Parameters" at the ***beginning*** and ***end*** of each leg of multiple planet trajectories. The C_3 (or energy) parameter would be calculated as a result of running SPARC to get the conic trajectories between two planets. A plot of the C_3 parameter versus time of flight (T_F) between two planets produces some numerically ill-behaved curves similar to those shown in Figure 2 below:

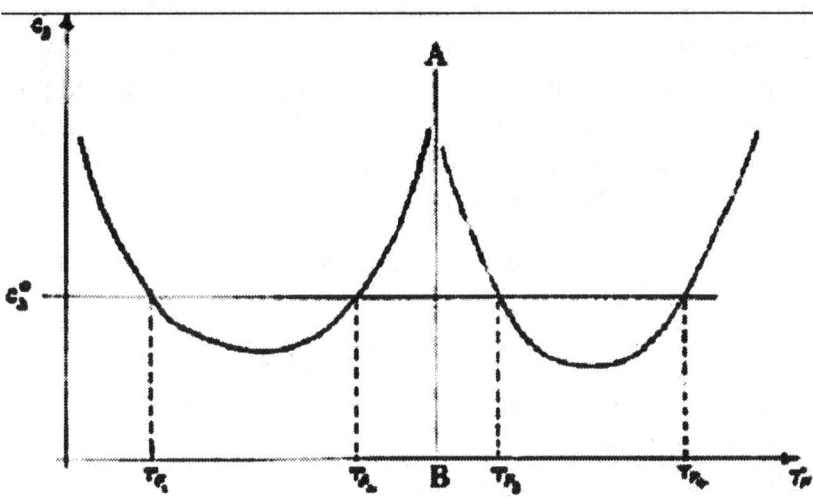

Figure 2

These C_3 vs. T_F curves were particularly ill-behaved where they become somewhat asymptotic to the vertical line segment AB. Going from one planet to the next required finding values of T_F for a particular value of C_3. A given value of C_3 would result in up to four different times of flight within one synodic period. When running SPARC a conic trajectory must be run for each one of the times of flight. This means that the grand tour trajectories grow in a tree like structure. At each planet there may be multiple times of flight for which conic trajectories must be run. The tree like growth creates a complicated bookkeeping problem of keeping track of each leg and times of flight (T_F) that go with each trajectory leg.

To find the values of T_F corresponding to a particular value of C_3 was accomplished by using a Newton-Raphson type iterative procedure. The Newton-Raphson iterative procedure was developed at TRW Systems, and this procedure was called a One Dimensional Search routine. The One Dimensional Search subroutine allows the user to specify the desired slope of the dependant variable when the solution is found. This feature was very important in solving for desired values of C_3 with respect to T_F. Figure 2 is an example of a plot of C_3 vs T_F. It shows that there are two concave-up curves on each side of line segment AB. Since the C_3 curve was so ill-behaved near line segment AB we decided that it would be beneficial to find the minimum value of C_3 on each of the two branches of the curve (that is, the branch on the left side of line segment AB and the branch on the right side of line segment AB). Finding the two minimum values of C_3 was useful in restricting the interval used to search for the values of C_3 corresponding to the positive and negative slopes of the curve. In finding the minimum values of C_3 we divided the interval using the Leonardo Fibbonacci (a 12th century mathematician) number to minimize the number of function evaluations; we assumed that the function was unimodal (i.e. the function had only one minimum value in the interval of interest).

After Ron Richard left JPL, Gary Flandro started to use SPARC to determine real Grand Tour Trajectory possibilities that were being planned at JPL for the National Astronautics and Space Administration (NASA). In order to make SPARC run faster, Gary developed a numerical ephemeris that was much faster than using the JPL Ephemeris Tapes for positions and velocities of planets. I programmed the new numerical ephemeris and put it in SPARC as an option. We could then make runs using either the numerical ephemeris or the JPL Ephemeris tapes. I enjoyed working with Gary and helping him to get the information he needed to proceed with the studies he was doing.

There is a statement on page 71 of Swift's book "Voyager Tales" where Dr. Gary Flandro said the following:

> "Before leaving JPL, I enjoyed working briefly with Al Joseph, in the design of a computer program that could be used to accomplish the analysis of multiplanet missions in a more automated fashion. Al produced an excellent program that enabled the complex searching procedures to be carried out without the tedious handwork that I had used in the first Grand Tour study. Later versions of Al Joseph's program were used, I believe, in the detailed Voyager trajectory design work. Such efforts as these in the late 1960s demonstrated to me that there would soon be an explosion of interest in outer planet exploration at JPL."

I would like to refer you to pages 62 and 63 of Swift's book "Voyager Tales" where Dr. Gary Flandro discusses who invented the concept of the gravity assist. I want to warn you of the dangers in taking information out of context from another publication. Following are some of the things that Dr. Flandro said:

1. I was already familiar with the idea of using intermediate planet orbit perturbations in the design of interplanetary trajectories. Most of my practical knowledge of the technique came from study of the works of Krafft Ehricke. His textbook, *Space Flight (Vol. II, Dynamics)* published in 1962, presented a comprehensive discussion of the application of the energy gain or lose speed relative to the sun.

2. He (Krafft Ehricke) saw with great clarity the benefit of using the massive planet Jupiter as an energy source in the exploration of the outer planets.

3. I was of course aware of Michael Minovitch's efforts, and it was my impression that he had been concentrating on the inner planets-Mars and Venus round trips. He had just published a JPL technical report that described solar system escape, out-of-ecliptic, and close solar probe missions utilizing Jupiter swingbys. In fact these had already been proposed by Ehricke. On pages 1107-1109 of Space Flight, Vol. II are detailed diagrams showing the trajectory designs just mentioned.

4. Minovitch later claimed he had invented the concept of gravity assist, or what he termed "gravity thrust," but in fact, I believe Ehricke should be given careful consideration if an inventor is to be identified.

5. In reality, the basic ideas behind gravity assist were known as far back as the 1800s. Astronomers (Leverrier, the discoverer of Neptune among them) interested in comet motions were quite familiar with the gravitational modification of the orbit of a comet or asteroid when passing close to a massive planet like Jupiter.

I am not trying to draw any conclusions or sway any opinions about who discovered the gravity assist by referencing the above discussion by Dr. Flandro. Again, I encourage you to read the above discussion by Dr. Flandro in Swift's "Voyager Tales." I also encourage you to read Dr. Minovitch's articles which appear in Volumes 19 and 20, AAS History Series.

The gravity assist is an important part of a multiple planet or grand tour computer program, but it is only a small part of the overall capability. By this I mean that the gravity assist by itself is not a grand tour computer program. Some of the other so called grand tour trajectory computer programs in the 1960s were programs that required multiple computer runs. SPARC was different from these programs in that it could run multiple planet or grand tour trajectories over a *range of launch dates* in a *single computer run.*

In 1967, after I had completed the SPARC program, I was appointed "Cognizant Programmer for the Double Precision Trajectory Program (DPTRAJ)," see Attachment B. I was still the cognizant programmer for SPARC. After I became the cognizant programmer, for DPTRAJ, JPL decided to switch from the IBM 7094 computers to the UNIVAC 1108 computers. DPTRAJ was an extremely large program and was second in size only to the "Double Precision Orbit Determination Program (DPODP)." DPTRAJ and DPODP had had large groups of programmers working on them for decades, so you can imagine how large they were. DPTRAJ was a subprogram to DPODP and together they may have been the largest computer program in the world. There were many JPL projects that depended on a timely transition of DPTRAJ from the IBM 7094s to the UNIVAC 1108s. John Strand and I decided that waiting to start the conversion to the UNIVAC 1108 was not going to help anything. No one had started to convert any programs to the UNIVAC, so we decided that it was to our advantage to start converting DPTRAJ while there was virtually no activity on the UNIVAC. We started by compiling hundreds of subroutines and trying to resolve the thousands of error messages that resulted from the compilations. Virtually every subroutine had some error messages. John and I worked like two mad men, and eventually got DPTRAJ working on the UNIVAC so that we would get essentially identical results on the two types of computers. John worked with the engineers to resolve many technical engineering issues that were increasingly more complex to analyze (see Attachment C). The reason that I decided to write about DPTRAJ is because DPTRAJ and DPODP were needed to do the precise runs for the upcoming Grand Tour Trajectories. We succeeded in converting the first JPL program to the UNIVAC 1108, and as I stated earlier this was the second largest program at JPL.

Skip Newhall wrote a program for the IBM 7094 computer called JPTRAJ. JPTRAJ was a monitor program that could run independent programs like subroutines and pass information from program to program. JPTRAJ used a language that could specify which quantities should be passed to programs that would run later. We needed this capability on the UNIVAC 1108 computers. Since JPTRAJ was written in IBM assembly language it could not be directly converted to the UNIVAC 1108. I spoke with the UNIVAC representatives and found that the UNIVAC FORTRAN was a far more powerful programming language than FORTRAN on the IBM computer. After getting the essential technical information from the UNIVAC representatives, I started to create the Trajectory Monitor (TRAM) computer program in FORTRAN. The main feature of the UNIVAC FORTRAN that made it possible to write TRAM in FORTRAN was the ability to start a computer program from a FORTRAN subroutine. TRAM like JPTRAJ had a language that specified which programs would be run and what information would be passed from program to program. The NAMELIST feature in FORTRAN generates a symbol table where the Binary Coded Decimal (BCD) names are followed by the address in the computer where the quantities are stored. The NAMELIST symbol table was used to extract and insert quantities that were to be passed or received by programs. TRAM was another innovative program that I created at JPL. The 1960s was a period where I was extremely creative.

After we developed TRAM, Robert Mitchell started to run many of the precise trajectories needed for the Voyager mission. As stated earlier these computer runs consisted of running DPODP, DPTRAJ, the Mid-Course Maneuver Program, etc. These runs would take several hours, so I always ran them overnight. I think that Robert Mitchell made as much of a contribution to the success of the Voyager mission as anyone who worked on the project. He may or may not have received awards for his work, but he certainly deserved some. I knew of this because I was personally submitting the computer runs for him.

I have given a considerable amount of thought to why I was very creative during the 1960s and 1970s. When the aerospace companies were using the IBM and UNIVAC computers during the 1960s and 1970s, the programming languages were not overly complex and almost anything could be done with the assembly languages, FORTRAN, PL1, etc. After this period the academic world began influencing the programming languages and languages like C, ADA, Algol, etc. were introduced. The academic world convinced the aerospace companies to switch to the more complex languages, thinking this would cause better computer programs to be written. The problem with programming was not the languages, but it was that the schools were not teaching students how to structure large computer programs properly. Programmers are still producing bad software with the new languages. I was lucky to be taught how to program computers by some extraordinary experts. Charles Haberman, Paul Steiner, and Robert Schreiner of Northrop Corporation taught me how to program in 1957. I am thankful that I met George Timpson of Space Technology Laboratories in 1959, who was one of the best at structuring computer programs. George Timpson was the person that I met who knew the advantages of putting the control logic of a program in the main program rather than spreading it throughout the program. A large computer program that has its control logic spread throughout the program can become so complex that no human being can maintain it or conveniently change it. Many large computer programs that have been developed by spreading control logic throughout the program can never be converted; many times you are better off starting from scratch to do the program again. The complex nested logic of programs when spread throughout the program, is extremely difficult to unravel. Centralizing the control logic in the main program makes it easier for everyone to understand the program. This simplifies the maintenance of the program, it simplifies making changes to the program, and you can convert the program to another computer without too much trouble. I think that the spreading of control logic throughout programs begins when students are taught programming in college or elsewhere. You can spread control logic throughout a small program and get away with it, but you can't do that in very large computer programs. I am amazed at how long the concept of centralizing the control logic in the main program has been around, and I am equally amazed at how few programmers know about it.

There were many programmers at JPL during the 1960s and 1970s that made significant contributions to the success of projects. The programmers have not received a lot of recognition and they have not received any awards that I know of, but they have the satisfaction of knowing what their contributions were.

Over the years I have wondered if I would have been able to create a computer program like SPARC if I had not known about recursive subroutines. I have never been able to think of another reasonable way to do SPARC other than the way I did it. After I finished SPARC, JPL showed their appreciation by giving me a permit to park on the laboratory premises and JPL treated me as a special person for the rest of the ten years that I worked there.

Attachment A is a letter of commendation from Ronald Richard for the work that I did on SPARC. Attachment D has a list of some related publications. Attachment E has some acknowledgments for work on DPTRAJ.

I would like to thank John Strand for encouraging me to write these notes. I don't think I would have done it without him pushing me to do it.

ATTACHMENT A

JET PROPULSION LABORATORY
MEMORANDUM

INTEROFFICE

312.4-544

September 14, 1966

TO: Distribution

FROM: R. J. Richard

SUBJECT: Commendation of A. E. Joseph for His Work on SPARC

DISTRIBUTION: E. Cutting, G. Gianopulos, T. W. Hamilton, W. Melbourne,

R. Y. Roth, J. F. Scott, F. M. Sturms

SPARC, JPL's multiple planet heliocentric conic trajectory program, is nearing completion. When completed, SPARC will have all of the capabilities that have been requested. That this is so is due in very large measure to Al Joseph. The requests for capabilities were both extensive and demanding, and, although the ability to implement all of the requests was in doubt several times, a highly satisfactory solution was always produced.

SPARC could not have been developed with the cognizant engineer only writing RFP's, and the cognizant programmer only programming. Several unforeseen and difficult problems cropped up, as, for example, in the important automatic searching methods for determining multiple planet trajectories. Some of these problems required extensive discussions to discover the causes and cures of the difficulties, and a continuous dialogue on all aspects of the program was carried on from the beginning. As a result, I have gotten to know Al quite well, and I can truly say that working with him has been a very pleasurable experience. I have only words of praise for his work, and pleasure with having worked with him.

RJR:mjf

ATTACHMENT B

JET PROPULSION LABORATORY INTEROFFICE
MEMORANDUM

TO: Distribution Date: May 9, 1967

FROM: G. N. Gianopulos REFR: 315.10/359

SUBJECT: Appointment of Alva Joseph to the Position of

Cognizant Programmer for DPTRAJ

Effectively immediately Mr. Alva Joseph is appointed Cognizant Programmer for the Double Precision Trajectory Program in the Trajectory Group, Section 315. Mr. Joseph has had the responsibility for the design and development of the Space Research Conic Program (SPARC). He will continue some responsibilities in this area to effect an orderly phase out. However, with this appointment, he is transferred to the Trajectory Subgroup of the Trajectory Group.

GNG:bje

<u>Distribution</u>

R. Jirka

Section 315 Engineers

Section 312 Engineers

ATTACHMENT C

April 17, 1968

Dr. Robert Rector

Informatics Incorporated

5430 Van Nuys Blvd.

Sherman Oaks, California 91401

Dear Dr. Rector:

On April 16, 1968, JPL made a presentation of its Double Precision Trajectory Program (DPTRAJ) to Marshall Space Flight Center and NASA Manned Spacecraft Center. Mr. John Strand did a very outstanding job in his part of the presentation. This was another example of the type of work John has done since he has been working on DPTRAJ and we would like to express our appreciation.

                                      Very truly yours,

                                      Alva E. Joseph

                                      Sr. Engineer

                                      Trajectory Group

AEJ:trb

ATTACHMENT D

LIST OF PUBLICATIONS BY ALVA JOSEPH

Joseph, A. E., "Timing and Resolution Program for the Combined Strategic Missile Forces: 211.5B Program Operations Manual," TRW Classified Document number 30068-6236-TE-00, April, 1977.

Joseph, A. E., "Timing and Resolution Program for the Combined Strategic Missile Forces: 211.5A-1 Critical Design Review," TRW Classified document number 28044-6282-TE-00, July, 1976.

Joseph, A. E., "Timing and Resolution Program for the Combined Strategic Missile Forces: Program Operations Manual," TRW Classified document number 28044-6830-TE-00, February, 1976.

Joseph, A. E., "Timing and Resolution Program for the Combined Strategic Missile Forces: 211.6 Critical Design Review," TRW Classified document number 28044-6329-TE-00, December, 1975.

Joseph, A. E., "Timing and Resolution Program for the Combined Strategic Missile Forces: Programmer's Manual," TRW Classified document number S/N 28044, November, 1975.

Joseph, A. E., "Design for the Mars Orbit Insertion Operations Program," JPL 620-104, March, 1974.

Joseph, A. E., and Capen, E., "Software Requirements for the Mars Orbit Insertion Operations Program," JPL 620-4, March, 1973.

Joseph, A. E., and Capen, E., "Orbit Insertion Design and Analysis Program," JPL 915.31/489, June 29, 1972.

Joseph, A. E., "JPL Double Precision Trajectory Program," NASA Technical Brief No. 71-10390, October, 1971.

Joseph, A. E., "Double Precision Analytic Partials Program," JPL IOM 315.31/237, January 29, 1971.

Joseph, A. E., "Time Data Sequential Processor," NASA Technical Brief No. 70-10720, December, 1970.

Joseph, A. E., "User's Guide for the JPL Double Precision Trajectory Program," JPL 900-317, October 17, 1970.

Joseph, A. E., "JPL Spacecraft Prediction System," JPL 900-337, December 30, 1969.

Joseph, A. E., "Multiple Planet Conic Trajectory Program," JPL 900-???, October 17, 1969.

Joseph, A. E., and Strand, J., "Supplement to the Double Precision Trajectory Program's Documentation," JPL 900-153, May 31, 1968.

Joseph, A. E., "Reverse Solution to Lambert's Theorem," JPL IOM 315.21/112, May 25, 1965.

Joseph, A. E., "Design for a Multiple Planet Conic Trajectory Program" JPL IOM 315.21/122, May 25, 1965.

ATTACHMENT E

Acknowledgment

The author wishes to express his appreciation to Mr. Francis Sturms, group supervisor of the Trajectories and Performance Group, and to Mr. Ahmad Khatib, cognizant engineer of DPTRAJ, both of JPL Systems Analysis Section (Section 392), for their efforts and contributions to this paper. In addition, he would like to extend appreciation to Mr. Fred Lesh, supervisor of the Trajectory Group, to Mr. Alva Joseph, cognizant programmer, to Mr. Joseph Witt and Dr. Thomas Talbot, all of the JPL Flight Operations and Deep Space Network Programming Section (Section 315), and to Mr. John Strand of Informatics for providing valuable insights into the computer program DPTRAJ.

Note

The above acknowledgment is from the following document: JPL Technical Memorandum 33-451, Design and Implementation of Models for the Double Precision Trajectory Program (DPTRAJ), by Gerd W. Spier, April 15, 1971.

APPENDIX II

The following unedited short story was first published in 1953 in THE FENCE POST. THE FENCE POST was the monthly newsletter for my Baldwin Park Boy Scout Troup. Thank you to my Den mother, Aleta Hitchcock, for saving the only copy. The story was printed in installments over a five-month period in 1953. (I guess my head was in the clouds even when I was twelve years old.)

THE EARTH'S END

John Richard Strand, age 12

Chapter One - - "The Motor"

Steve Walker walked merrily along to his laboratory. He was thinking about his new motor he had invented. He said to himself, "I hope I get my new car motor done before Professor Barkins gets here to inspect it. If I don't have it finished he will probably go away and it will be another month before I get to show it to him again." He came to his lab, walked up the stairs, opened the door, and walked over to his small bench which held a souped up car motor and some other parts. He worked on seriously until he glanced at his watch; it read 5:00. "Oh my goodness, supper will be just about ready, I had better eat and get a good nights sleep."

In the morning he went back to work and by 2:00 he was finished. He said to himself, "I hope this thing works. We'll know in a minute I'll push the starter button. Something must be the matter, it won't work. All of that work for nothing. What will I tell the professor when he comes?"

When the professor got there Steve told him the bad news. The professor said "I might as well have a look at it, after all of that work."

Steve showed him the motor. The professor said, "Very good, but if I put this part here it might work. All right now, start the motor." The motor made a little noise and then came to a stop.

"I've got an idea," said the professor, "but I have to have a lab. Mine is clear across the country"

"You can use mine," said Steve, "Say, what's your idea?"

"This motor of yours has given me an idea about a rocket ship that is driven by atomic radiation. This rocket will take a lot of testing, but I'm sure it will work, at least I hope it will."

Chapter Two - - The Rocket

(Steve Walker had just completed his new car motor for professor Barkins to see. The car motor was a flop but the professor had an idea for a new rocket.)

For the next three months Steve and the professor worked on their model rocket ship. On Jan. 5 their first test rocket was finished. It was a small model about 5 feet long. The real rocket would be 72 feet long. The next day it was tested. Eggs were put in it for human beings.

When it started it went off of the ground; it spun and fell to the ground breaking all of the eggs. Steve and the professor felt dismayed. That night the professor said, "Do you think we should try it again?"

"Well," said Steve, "If we put some rockets on the sides it might not spin."

"Good idea!" exclaimed the professor, "We might have a chance of succeeding after all."

So again they set to work. They salvaged some of the parts from the old rocket and the new one was finished in two and one half months. After they had tested the rocket the report read: "It flies okay but when it lands it breaks up so we'll have to fix a suitable landing gear."

Two days later a strange thing happened while the professor was testing with some electricity he accidentally touched a piece of metal in such a way that it made it very light. The professor then set out to make a landing gear out of the lightened metal. Finally after much strenuous work it was finally finished.

Steve and the professor showed the plans to Washington DC. They told them that if the Hall of Science would approve they would supply the men and materials to build the ship and to train the men for the trip. The Hall of Science approved. So the United States sponsored the trip.

Four years later the ship was finished and five men plus Steve and the professor were ready for the take-off. They were Captain Jefferson, the pilot; Steve was navigator; Captain Land was rocket engineer and the rest were trained in everything so in case anything happened to any of the other men they could take over. The night before the flight each of the men was given a sleeping pill to get a good nights sleep.

In the morning all of the men were on hand for the take-off. Then the order came "Prepare for take-off." The men turned on the oxygen and fastened their safety belts to prepare for the blackout period.

$$10 - 9 - 8 - 7 - 6 - 5 - 4 - 3 - 2 - 1 - ZERO!$$

Chapter Three - - "Breakdown"

(Steve had invented a car motor which was a flop. Afterward professor Barkins and Steve invent a rocket from the motor. They have just blasted off....) The rest of the story will be in Steve Walker's own words.

The blackout period was just over. I was just coming to when I heard professor Barkins say, "Something's wrong. We're supposed to be going 2000 miles an hour, instead we're going 5000 miles per hour. That's faster than the things supposed to go!" "The speedometer keeps climbing, 6000, 7000, 10,000. Finally it will not go any farther!" "Shut off the rockets!" shouted the professor.

"I can't," shouted Captain Land. "The throttle is broken. I'll go out on the hull and fix it before we go so fast that we burn up." Captain Land went out on the hull of the ship, but he forgot his gravity boots. His guide rope broke and he floated away from the ship. Seeing this I jumped into my suit and rushed out on the hull and threw a rope, but he had floated far away. Then I had an idea. I went back into the rocket and got a miniature rocket. I fastened it to my back.

I shot off to the end of my guideline and threw my extra line to Captain Land, but it missed. I gathered it up. I "have" to make this one, my rocket is getting weak. Easy...easy...Made it! Now to get started back.

Just then my rocket went out!!

Chapter Four - - "Strange Planet"

(Captain Land went out on the hull of the rocket but floated away. Steve went to rescue him and managed to throw an extra line to him. But, Steve's miniature rocket on his back went out.)

Due to the quick thinking of professor Barkins I was pulled safely back to the ship with a rope he threw me.

Inside the professor was looking at the controls then said to me. "I think I've found out what happened. Out in space there is no air to hold us back like in the Earth's atmosphere, so out in space we go many times faster."

"Well then," said Captain Land, "We must be in a different star galaxy."

"That's right," said professor Barkins, "I'll set the controls down a little."

Day after day we went through the strange star galaxy. Each night we studied the stars. Then one night I sighted a small planet and set our course for it. Five days later we landed on the planet where we found a demolished city. Upon landing we found the atmosphere full of atomic radiation, so we put on our atomic radiation decomposers. We got out of the ship and we all stared at a once beautiful city, now a mess of rubble. The professor told me later it must have been an atomic explosion many years ago.

We went into a half standing building that looked like an old church. In the corner was a pile of books. We opened one and found that the lettering had turned to dust because it was so old. When we opened it the dust blew away.

"Hey!" shouted the professor, "Here's one that's not quite so old, but look at the scribbling. It looks like the old type of dialogue used by the Cave men." He read a few lines and turned around to me and said, "This book tells all about the people. It seems that they were giants 10 or 11 feet tall."

I suggested to the professor that we should look around. We went into the next room and there on the fireplace were ashes not more than one day old!!!

Chapter Five - - "Giants"

(Having landed on a strange planet, the crew has discovered that it was a land of giants that were supposed to have been destroyed by an atomic blast years before. As last month's segment closed, Steve and the professor had just discovered day-old ashes in a fireplace.)

"Let's get out of here!" said Captain Land. "I sure wouldn't want to meet one of those guys in a dark alley."

"Let's hurry," said the professor. "I'm sure I saw a shadow moving behind that rock."

Just then a stone ax came flying across the room and hit one of the men in the head and killed him instantly. We left him there because now it's every man for himself. We got in the ship and 5 giants started coming towards us.

"Wow!" shouted Captain Land, "Look at those meat choppers. They must be 3 feet long."

"Hey!" I shouted, "They're going to ruin our ship. We had better shoot them with our detenose rocket."

"Diameter 4" - "Circular 16" - "Set on" "5 - 4 - 3 - 2 - 1 - 0 - FIRE!"

A spine cracking sound was heard and one the giants fell while the others scattered.

"Bring that guy in. I want to see how that guy lives in this radiation." said the professor. We sure did have a hard time bringing him in because he was so big.

"We sure could use him on our basketball team," mused Steve. "I sure wish I was back home right now. Why did I come on this crazy flight anyhow? Let's dump this overgrown caveman overboard and head for home. This place gives me the creeps."

"Rodger," said professor Barkins, "Get ready for the knockout spell."

When we came to, the navigator said, "We are now in the Opal galaxy. In a few weeks we'll be in our own galaxy."

All of these good thoughts were ended by the clank of a meteorite. Our rockets went OUT!

Chapter Six - - "Earth's End"

(As Steve and the remaining crew headed home, a meteorite knocked out the ship's rockets)

We sent Captain Howard out to fix the rockets and that's when it happened. He must have fixed the rockets too good, for when they were fixed the rockets blasted off and burned him to a crisp. We had forgotten to turn them off.

"Well," said the professor, "That's two men we've lost on this crazy flight. I wonder if it was worth it?"

On and on we flew and finally we came to our own star galaxy. Finally we came to the thin layer of air that surrounds the Earth called the atmosphere. Soon we passed that boundary. It must have been the exposed atomic radiation that our ship had left when we were leaving the earth's atmosphere - - - the whole earth was just a bunch of ashes with a few blots of fire here and there.

Upon our blast-off the ship must have activated the atoms inside the Earth and started the world-wide fire and blew up. I don't know how it happened, but Captain Jefferson and I were the only survivors of the ship. How long could we last, was the question, since the Earth's oxygen was all consumed by the fire and my suit only had one hours supply. Then Captain Jefferson died of a head concussion.

So I, Steve Walker, was the only person on Earth. My air's getting thin, so I will have to end my story. But maybe in a few million years the Earth may come about again. But for the Earth's sake, I hope nobody will ever get enough atomic radiation to set off all of the atoms in the middle of the Earth again.

So, goodbye . . A H

              H

                        H

                                  H

                                          The End

APPENDIX III

NON-TILLAGE AGRICULTURE – John R. Strand

By its name, agriculture seems to indicate something beyond science; i.e. culture. I possibly became aware of non-tillage as a result of reading an article by Ruth Stout in <u>Organic Gardening</u> magazine or in a book she wrote called GARDENING WITHOUT WORK. She was a gardener who lived back east and wrote about her experiences. She simply and clearly explained that you don't need to dig up your garden in order to grow vegetables. Saving all that back breaking work and increasing the fertility of the soil was her clear and simple message. She used a mulching system that spread her organic refuse on top of the soil rather than burying it like a moldboard plow does. "Make your garden your compost pile" was her slogan. This method saved soil moisture, diminished the need for weeding, improved the soil structure and seemed to reduce labor. The top three or four inches of her garden were the focus of her attention.

My first impression was that it was a fanciful and theoretically interesting idea. But often I thought that some of the ideas expressed in the magazine sounded exaggerated. (I will never forget the one about the man who made manure tea in a thousand gallon tank and stirred it with his outboard motor). Some of the homespun techniques sounded like more work than traditional methods although clearly the gardeners were having fun. I have never wasted any beer on snails. My system was to drink the beer, then go out late at night with a flashlight and shovel, a much more effective method. I have observed snail herds of over a hundred trying to cross my lawn.

Gardening is a little like fishing; people develop their own method that works well for them. I would never consider any contraption for my compost heap. As Ms. Stout explained, it is unnecessary, nothing in the garden needs to be removed. Garbage, leaves, grass clippings or the neighbors' cow manure can be placed somewhere in the garden. The more "rank" materials can be pushed under the compost. I recommend you keep meat scraps out of your garden. Dog and cat litter should go elsewhere because of parasites. It matters not what season or what you are growing; there is always a place in the garden for everything organic.

My great-grandparents had settled in the northwest corner of Minnesota, perhaps because it reminded them of their homeland, Sweden. Virtually everyone was a farmer. My dad used a team of horses to plow acres of land for potatoes, wheat and corn. This is heavy work especially at threshing time. The soil in this part of Minnesota looks as black as coal in many places. It is laid down by the Red River and is considered to be the most fertile soil on earth. The rain and cold are the most important factors in this area. Plow it or disc it; the plants will grow if they have moisture and warmth. Plowing was a critical part of growing the crops that supported my family. Using Stout's method seemed diametrically opposed to these principles and may not necessarily be applicable where the land is very fertile. Because her garden was small, she may have had more time to tinker with it. But, with acres and acres of land, gardening techniques like this should be impossibly inefficient. I did still wonder about larger farm applications of non-tillage. If Stout was correct, couldn't these principles be used in some way on larger modern agricultural farms? After all plants are plants.

I was living on forty acres in Humboldt County, in California's outback. I decided to give non-tillage a try. One problem was that by the time I had read her articles I had already plowed up a natural meadow. I believe I was the first person to plow the sod at that location. The heavy clay was subsequently loosened up by over a foot of compost. My plow was a Troy Built and a shovel. What a test of man versus earth that was! A neighboring homesteader found a blue vein in the heavy clay near my meadow and used it to make pottery.

Hmmm, if I was to play by the rules then the shovel should be put away. Wrong again. I did notice that if you do not have unlimited mulch available because of your improvidence you may need to occasionally pull or chop up a weed. I now regard pulling weeds as punishment for ineptness in obtaining compost. As sometimes happens my mulching regimen slips. The most efficient way to weed is to pull them when they are the smallest. If you let them grow to two feet you may need gloves and a shovel. Weeds that size also re-seed themselves, making next season more difficult. To plow and bury the weed seeds only allows them to come back in future seasons.

Stout's method reduces, but does not eliminate the need for weeding. In a laboratory setting I would assume that mulching could eliminate all weeding. The imprecise real world usually requires some toleration for weeds depending on how you persevere with your mulches. Farmland size crops can also tolerate some competing weeds.

Stout also suggested some more complex advantages to non-tillage. Tillage destroys the soil structure. Structure is what helps air get to organic matter and make food for the vegetables. Food for the vegetables comes originally from the mulches or from residual nutrients dissolved within the soil. The whole picture as she described it was of a self-sustaining rotation involving organic matter, air, water, worms, legumes, nodules, bacteria and nutrients. It was simple and complex at the same time. Some mechanized equipment is unavoidable on a modern farm; on the other hand anything that compacts is the enemy of the soil. A plow sole is a compacted layer that prevents roots from reaching lower levels of soil. In fact, with poor soil the plant roots may be located almost exclusively in the upper mulch region.

Through the years I gradually perfected my technique. I no longer plowed the earth. The tiller became useless to me so I sold it. I did work the ground with the shovel occasionally. This was a quick chopping of the upper two or three inches because the spring weeds had gotten out of hand. This could have been avoided with winter mulching, but the weeds themselves were a good source of organic matter. They were just like a vetch cover crop. There always seemed to be exceptions to every rule.

I actually harvested a hundred pounds of carrots from a six-foot by six-foot square section using my clump method rather than rows. One of my kale plants lasted three years and produced over a hundred pounds of kale.

I continued to wonder about modern agriculture in terms of non-tillage. By now I had read a book published in the forties; "Plowman's Folly" by Edward Faulkner. This single book began to motivate me to spread the word like never before. I don't believe a single book at that time ever stated so clearly the principles that Stout preached. After reading "Plowman's Folly" I believed everyone would have to agree that non-tillage demanded a scientific investigation.

The Midwest was tamed by the plow. The rich soil and rain brought out ample bounty. Gradually tillage practices weakened the soil. Drought and wind created the dust storms of the "GRAPES OF WRATH."

Tillage is anathema to soil preservation. Tillage is a short cut to agricultural success. Our early pioneers needed immediate sustenance. They were justified in what they did at the time. We plow because we have always plowed. My grandfather did it, my father did it, I did it, and my son shall do it.

This is the legacy of the plow and it was still in full force in the 1970's. Because of my GI bill benefits and my love of gardening I attended College of the Redwoods in the early 1970's and took five agriculture courses. Mr. Regli and Mr. Richter were important influences helping me understand agricultural science. I continued to ask why weren't the marvelous results obtained in Stout's garden applicable to large-scale modern agriculture? I was baffled. It was as if Stout and "Plowman's Folly" were irrelevant. I decided to leave no stone unturned. I wanted an answer. Mr. Richter was very patient and seemed to agree with each point that I brought up. I still wondered why the world at large seemed to ignore it. He had no rational for this. I believe I was a bit of a nuisance in the Agricultural Department. I am sure I met with Mr. Richter over fifty times and he always welcomed our discussions.

This began a passion for traveling and asking persistent questions. I quickly connected with the agriculture agents in Humboldt County and found very little interest there. I also approached anyone who would listen.... at airports, on the highway, in grocery stores you name it.... I asked them why? I later spoke with Federal agricultural agents and got some positive responses but no definitive explanation. I also wrote to SCIENTIFIC AMERICAN appealing to them to publish something on the subject. After several years of crusading and speaking to about two hundred people I finally faced the inevitable question. Where would agricultural science be most revered?

The University of San Luis Obispo was the answer to that question. I arrived at the campus early one morning. I wanted to talk with the head of the Ag Dept. It happened to be the first day of the new semester. The chairman was a very congenial gentleman and gave me an hour of his very valuable time. I ran by him my conclusions from years of testing and studying. He listened intently. He told me that years earlier he had studied growing carrots in sawdust and understood my points. We arranged to meet again after lunch.

After giving a lot of thought to what I had said, he did a bit of research during lunch. He located a book published just that month. It was an exposition of non-tillage progress. Maybe people were paying attention after all. I expected this book to expound the virtues of organic gardening. Non-tillage was recognized as advantageous, but instead of using a plow, they were using weed killer! More chemicals! I did eventually understand why. The tinkering with mulches in Stout's garden would be facilitated by herbicides. The book agreed with Stout about leaving the harvest refuse on the ground or discing it into the surface.

I began to read all the agriculture magazines that I could lay my hands on. Gradually, month after month, I noticed references to non-tillage. Not directly though, as it seemed to be very controversial in the beginning. Eventually the floodgates opened. I noticed that the establishment was accepting non-tillage.

Soybeans were scattered after harvesting without any tillage. Stubble corn was actually left on the ground instead of being buried. Deep subsoil tillage was reduced; a friendlier and more passive attitude towards the soil was apparent. Mechanization was still important but the attitude of deeper is better was gone.

With modern weather irregularities and drought conditions, non-tillage would become even more important. I wonder what my Swedish ancestors would think of this?

Ruth Stout and Edward Faulkner were vindicated. Today the soil is barely disturbed after the sprouting weeds are killed with Roundup. The crop seed is then drilled into the soil. The coulter or harrow is only an inch wide and cultivates to the depth of one inch. No further cultivation is needed as that would bring to the surface more weed seeds to germinate. The soil is not tilled or disturbed until harvest. Midwestern crop yields increased, soil moisture improved and weed competition was reduced. Less mechanization meant less soil compaction and lower energy costs.

Genetically altered crops rather than chemicals may hold the key to the future of modern agriculture. I know my Swedish ancestors would definitely not understand this.

APPENDIX IV

CARDPUNCH - John R. Strand

Humboldt State University operated the last second-generation Honeywell computer in California. Computers and computing were not an important part of this university in 1980. This was the Redwood Curtain, as so many referred to our relative isolation.

The computer room was just as impressive as any you might have seen in the late 1960's. The large vacuum controlled tape drives and the staccato clattering of the high-speed printers caught your attention. There was still heard, even in 1980, the smooth whir of a card reader. There were the large color coordinated cabinets for the CPU electronics that dated this computer. Video monitors with keyboards along with other third generation improvements were an augmentation that had extended the life of this second-generation computer. I felt right at home with this card environment and started some research projects for the Physics Department.

A lower division physics experiment was consistently producing biased results. A piece of lead was immersed in liquid nitrogen until it attained the low temperature of the nitrogen. It was then placed in water and ice formed on the lead. The weight of the ice related to the energy change as the lead warmed up. There was consistently an unexplained error in the results. Dr. Lester Clendenning, chairman of the Physics Dept., asked me to help him investigate the source of this error.

Albert Einstein was of course known for his Theory of Relativity. It is not well known that he produced a model that predicts the increase in temperature of metals when energy is added. Not a startling breakthrough in fundamental physics like relativity, but interesting never the less. There are several competing models that use the degrees of freedom within atomic motion. In deriving Einstein's equations a numerical constant emerges which has the units of temperature. This is not the temperature of the metal, which could be any value up to the melting point. This is a <u>constant</u> that is specific to the metal (in our case lead) and has no relation to the temperature range of our experiment. This constant is called the Einstein Temperature and would be the focus of our investigation.

Lester handled the analysis which required specific heats obtained from the handbook of Chemistry and Physics. I would program in FORTRAN a numerical search that would calculate the Einstein Temperature for lead. It was hoped that our improved value would explain the discrepancy in the experiment.

We found a ten- percent difference between our result and the handbook value for the Einstein Temperature, but our laboratory discrepancy remained unresolved. Decades later Lester said the probable cause of this was that the lead we were using was alloyed with another metal. The real reward for our work was uncovering the most interesting programming bug I have ever heard of.

Honeywell made a fine second-generation computer. Only because of IBM's total dominance of the computer market in the late 1960's is this machine not better remembered. Higher maintenance costs, less addressable storage and slower processing speed made this a real dinosaur in 1980. Mr. Mild and Mr. Wilson, managers of the computer facilities, were delighted when they learned a new third generation Honeywell would soon arrive. The old computer would breath its last as a sledgehammer reduced it to scrap. Alas, my card programs were obsolete, so I rushed to complete the Einstein Temperature project.

To speed up the work I decided to copy my source deck so I could submit more jobs. I had noticed a problem with deck duplication. The duplicated decks were always flawed and I had to debug the program a second time to locate cards punched incorrectly. The IBM cardpunch I had worked with decades earlier had been significantly more reliable.

Mr. Wilson explained that the Honeywell cardpunch had a notorious endemic problem. Throughout the industry it was known to be unreliable. By 1980 cards were an archaic instrument of the past anyway. Later Mr. Wilson told me he had discovered the most incredible thing. A system-programming bug was responsible for the cardpunch malfunctions.

HSU maintained a complete listing of the source code for the Honeywell operating system. I had sparked some interest in the cardpunch and Mr. Wilson had casually examined the routines responsible for their execution. He noticed a timing loop that was not appropriate. Timing loops are repetitive instructions of no consequence, which slow the CPU. Upon further investigation he concluded this must have been a part of testing the cardpunch when it was first operational. Mechanical and computational compatibility is important. This appeared to be a study of the performance capabilities of the cardpunch. The timing loop would control the speed of the mechanical process. Unfortunately this was not removed when the original system was certified. The bug languished for the duration of this computers' operational life and was discovered the week the last example was put to the sledge. I found it fascinating but wondered how many more of these computers might have been sold if the cardpunch had been reliable?

ACKNOWLEDGMENTS

I want to mention Dr. David Swift author of VOYAGER TALES. David, a professor at the University of Hawaii, encouraged me to write this book. His help was invaluable and I am so happy to have met this distinguished man. Thank you, David, for your assistance and the confidence that you gave to me.

Thank you Dr. Donald C. Elder III, of Eastern New Mexico University, for wading through the drafts to locate the pearl.

The writing by Alva Joseph on the appendix is outstanding. Thanks Al for all the work you put into the true origin of the Grand Tour!

To Randy, Aaron, Austin, Perry and TJ at Panacore, thank you for the inspiration and encouragement.

Also, thank you to my wife Linnea, who typed and retyped again and again, while keeping me on track.

John Strand

INDEX

A

Abacus 61
Atomic clock 147
Avery Label 17, 25, 26, 31

B

Bambi 135
Beanie 1
Belden 148, 149
Boggs 73, 94
Boy Scout 8, 18, 19, 187

C

C++ 141
Calculus 49
Cartesian 46, 49
CLAM 147, 150
Comet 141
Congressional Award 89
Core dumps 123
Corning 151, 158, 160, 161, 163
Cowell 49
CPU 67, 68, 69, 99, 103, 197, 198
Crest Building 130
Croix De Guerre 113

D

Death Valley 73, 74, 82, 103, 134
Deep Space Net (DSN) 43, xvii
de Sitter 46
Diodes 83
Disneyland 6
Doppler 43, 47, 139, xvii

E

Earthquakes 147
Earthquake Affiliate Conference 147
Enke 49
Ephemeris Time 48
Euclidean 46
Euler 46, 49, 50
Everett 49

F

FASTRAJ 113, 131
FASTRAN drum 69
FORTRAN 31, 33, 43, 44, 51, 53, 61, 64, 65, 68, 99, 107, 108, 123, 129, 136, 141, 144, 163, 174, 197, xix

G

Galileo 37, 140, v, xv, xix
Gauss 43, 97, xvi, xix
Global positioning xix
Goddard xv
Godzilla 135
Golden Gate Bridge 103, 104
Goliath 78
Grand Tour 44, 45, 167, 168, 170, 172, 173
Gravity 46, 112

H

Hand checks 51
Hildebrandt 90
Honeywell 63, 197, 198
Hopalong Cassidy 1
Householder 47, 97
Huggy Boy 6

I

Industrial spying 162
Infeld 46
Informatics 44, 52, 69, 77, 78, 119, 135, 181, 185
Injection Conditions 94
Instantaneous maneuver 50

J

Jet Propulsion Laboratory (JPL) 167, xv
Jupiter 167, 172, 173

K

Keypunch 64
Kilroy 33

L

Lagrange 46
Lasers 131, 147
Leibnitz 52

M

Mariner 37, 43, 44, 57, 64, v
MASCONS 46
McBain 157, 163

Mercury 10, 113, xvi, xx
Microwave 139
Mid-course maneuver 44
Motor burn 46

N

NASA 32, 38, 43, 44, 51, 53, 66, 67, 100, 115, 139, 172, 181, 183, xv
National Numerical Analysis Society 51
Neptune 89, 167, 173
Newton 10, 43, 46, 50, 52, 53, 112, 115, 165, 171, v, xv, xvi, xix, xx

O

Oblateness 46

P

Pasadena Star News 57, 73, 74, 82
PASCAL 162
Phase change 93
PhD 44
Pickering 57, 113
Pixels 159
Pluto 115, xv

R

Range 43
Reichert-McBain 137, 157, 158, 161, 162, 163
Relativity 38, 46, 51, 112, 115, 123, 131, 197, xvi
Rube 32
Rubidium crystals 139

S

Saturn 32, 34, 115, 165, 167
Saturn booster 32
Schwartzshield 46
Scientific American 10, 145
Section 125, 179, 185
Secular forces 48
SFOF 64, xviii
Sputnik 31, 44, 90, v, xv, xix
Sun 44, 48, 93, xv

T

Taylor 49, 50, 143
Telemetry 43, 47

U

Uniscope 67, 68
Uranus 89, 167

V

Venus 113, 172
Viking 43, 131, 139, 140, v
Von Braun 34, xv

W

Warner-Lambert 163
Wilma 17, 18, 19

ABOUT THE AUTHOR

John's teenage year's could be best described as a true "American Graffiti." At seventeen he entered the Army and was trained as a medical laboratory technician. This was the beginning of a life-long passion for Science. Following the service Avery Label employed him as a chemist. John developed a patent and adhesive formulations that significantly contributed to Avery's explosive growth.

Having studied mathematics in college, he decided to join North American and worked on spacecraft trajectories for project Apollo. With computer and analysis skills now well established he started work for the Jet Propulsion Laboratory. This work produced the Navigation system used for projects Mariner, Viking, Voyager and beyond. His contribution to the world's knowledge of celestial mechanics resulted in a congressional award. This work, described in a Scientific American article, represented advances in the world of Physics, Mathematics and Computers, including a verification of Einstein's theory of Relativity.

As a Navy consultant he contributed to a project called VAST, which was a breakthrough in electronic component automation.

John's future plans are to promote preservation of the historical records of scientists associated with landing men on the moon and the exploration of the planets.

**Saturn's Rings
(8-25-1981)**

Courtesy: NASA Archives

www.ingramcontent.com/pod-product-compliance
Lightning Source LLC
Chambersburg PA
CBHW081112170526
45165CB00008B/2425